A PHOTOGRAPHIC GUIDE TO THE

BIRDS

OF JAPAN

AND NORTH-EAST ASIA

TADAO SHIMBA

Yale University Press
New Haven and London

Published 2007 in the United Kingdom by Christopher Helm, an imprint of A & C Black Publishers Ltd., and in the United States by Yale University Press.

Commissioning Editor: Nigel Redman
Project Editor: Jim Martin
Design: Wordstop Technologies (P) Ltd., Chennai, India

Printed and bound in China

Library of Congress Control Number: 2007931515
ISBN 978-0-300-13556-5 (pbk. : alk. paper)

A catalogue record for this book is available from the British Library.

The paper in this book meets the guidelines for permanence and durability of the Committee on Production Guidelines for Book Longevity of the Council on Library Resources.

10 9 8 7 6 5 4 3 2 1

130°
135°
140°
145°
45°
40°
35°
30°
Rebuntou
Rishiritou
Teuritou
HOKKAIDO
Daikoku-jima
Okushiritou
AOMORI
AKITA
IWATE
Hideshima
Sangan-jima
Tobishima
Awashima
YAMAGATA
Sadogashima
MIYAGI
Hegura-jima
NIIGATA
FUKUSHIMA
TOYAMA
TOCHIGI
GUNMA
Oki
ISHIKAWA
NAGANO
SAITAMA
IBARAKI
FUKUI
GIFU
TOKYO
JAPAN
YAMANASHI
TOTTORI
KYOTO
KANAGAWA
CHIBA
Mishima
SHIMANE
OKAYAMA
HYOGO
SHIGA
AICHI
SHIZUOKA
Oshima
Tushima
HIROSHIMA
OSAKA
MIE
Nijima
YAMAGUCHI
KAGAWA
NARA
Kouzushima
Miyake-jima
Iki
EHIME
TOKUSHIMA
Mikura-jima
Izu Islands
FUKUOKA
KOCHI
WAKAYAMA
SAGA
OITA
Hachijo-jima
NAGASAKI
KUMAMOTO
MIYAZAKI
KAGOSHIMA
Torishima
Tanegashima
Yakushima
0
500
1000 Kilometres

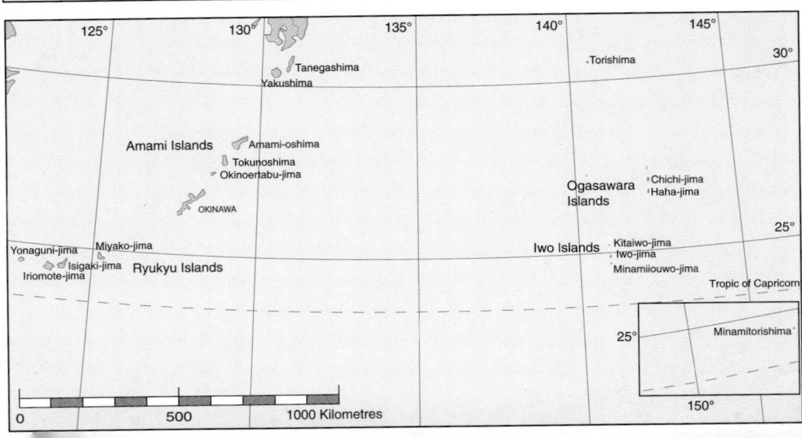
125°
130°
135°
140°
145°
30°
25°
Torishima
Tanegashima
Yakushima
Amami Islands
Amami-oshima
Tokunoshima
Okinoertabu-jima
OKINAWA
Ogasawara Islands
Chichi-jima
Haha-jima
Yonaguni-jima
Miyako-jima
Isigaki-jima
Iriomote-jima
Ryukyu Islands
Iwo Islands
Kitaiwo-jima
Iwo-jima
Minamiiouwo-jima
Tropic of Capricorn
25°
Minamitorishima
150°
0
500
1000 Kilometres

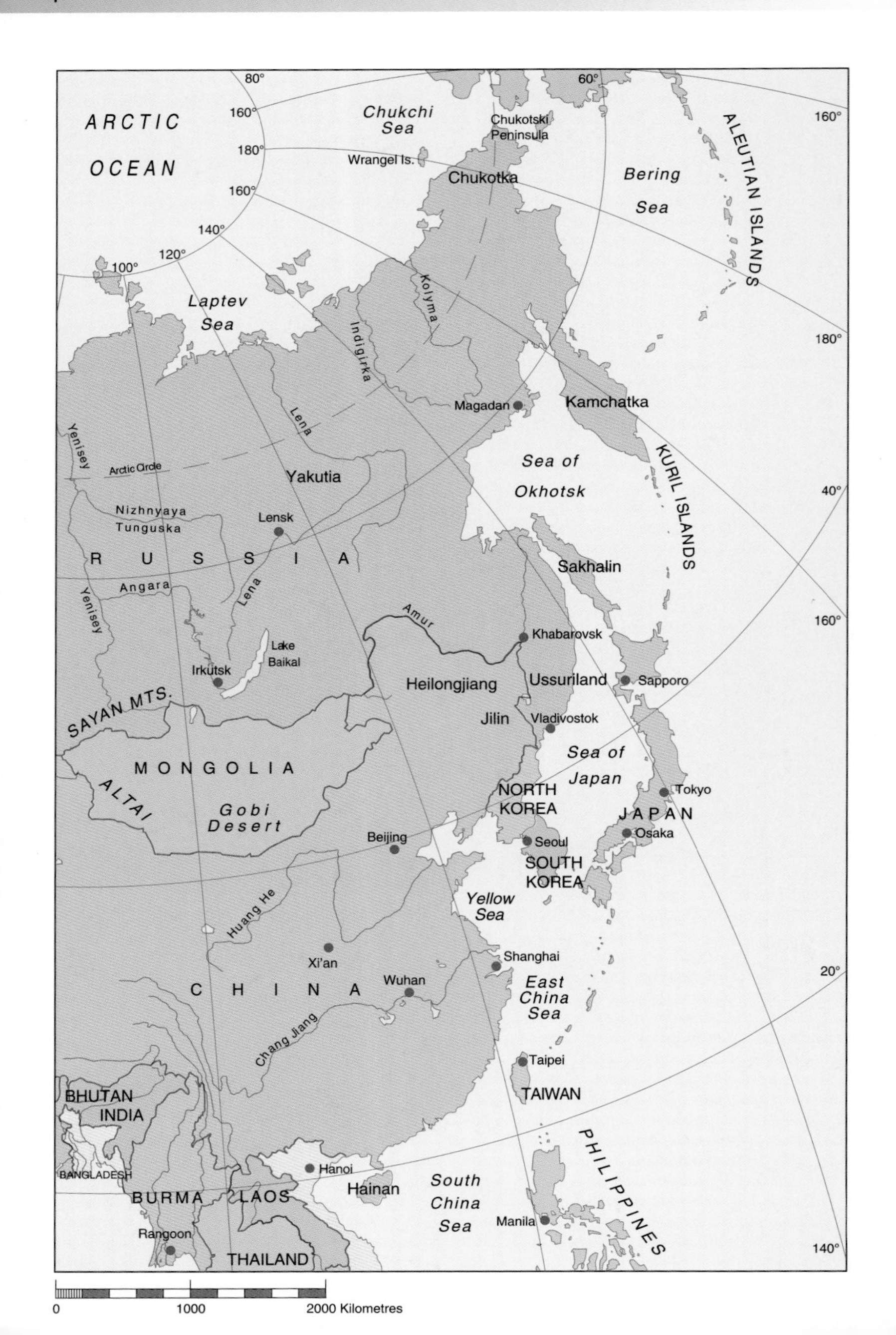
ARCTIC OCEAN
Chukchi Sea
Chukotski Peninsula
Wrangel Is.
Chukotka
Bering Sea
ALEUTIAN ISLANDS
Laptev Sea
Kolyma
Indigirka
Lena
Magadan
Kamchatka
Yenisey
Arctic Circle
Yakutia
Sea of Okhotsk
KURIL ISLANDS
Nizhnyaya Tunguska
Lensk
R U S S I A
Sakhalin
Angara
Amur
Khabarovsk
Lake Baikal
Irkutsk
Heilongjiang
Ussuriland
Sapporo
SAYAN MTS.
Jilin
Vladivostok
Sea of Japan
MONGOLIA
ALTAI
Gobi Desert
NORTH KOREA
Tokyo
JAPAN
Osaka
Beijing
Seoul
SOUTH KOREA
Yellow Sea
Huang He
Xi'an
Shanghai
East China Sea
C H I N A
Wuhan
Chang Jiang
Taipei
TAIWAN
BHUTAN
INDIA
PHILIPPINES
BANGLADESH
Hanoi
Hainan
South China Sea
BURMA
LAOS
Manila
Rangoon
THAILAND
80°
60°
160°
180°
160°
140°
120°
100°
160°
180°
40°
160°
20°
140°
0
1000
2000 Kilometres

CONTENTS

ACKNOWLEDGEMENTS

MANY FRIENDS SUPPORTED ME throughout the publication process of this book, and finally, with their help, I managed to make it happen. In particular, I appreciate the support received from Christopher Helm Publishers; Nigel Redman and Jim Martin spent time listening to my ideas, gave me useful advice and constructive comments, helped gather and locate a number of photographs for me, and were instrumental in the design and layout of the book. I also thank Ernest Garcia for his skilful editing.

Tokio Sugiyama, Akira Yamamoto and Hiroshi Kakimoto, who joined this project in its early stages, provided many wonderful photos and useful advice. Shigeo Ozawa and Mitsuo Imai, who have been my birdwatching friends for more than 30 years, gave me advice on the identification of birds and also provided me with photographs. This book would not have been realised without their cooperation.

In addition to Tokio, Akira, Hiroshi, Shigeo and Mitsuo, many other photographers have allowed us to use their work. I am especially grateful to Kazuhiro Kisaichi, Norio Yamagata, Manabu Totsuka, Himaru Iozawa, Kenji Takehara, Nobuo Shiraki, Teruaki Ishii, Yoshihito Goto, Yoshio Goto, Hyuntae Kim and Tomi Muukkonen, who all provided me with many valuable pictures.

Takaaki Inoue, a birding friend for more than 30 years, and Takefumi Abe of Bionomicslab Co. Ltd drew the distribution maps for each species. I thank them for their carrying out this task, which required a great deal of patience and skill.

Another birder friend, Chris Tzaros, patiently helped with the description for each species; Mitsuyo Matsumoto, Atle Van Olsen and Robert Edmunds provided me with useful assistance in communicating with an English-speaking readership.

And finally, I would like to thank my family, who supported me throughout the production of this book. I would like to give particular thanks to my wife Yukiko, who has always shown such understanding.

INTRODUCTION

THE MAIN OBJECTIVE OF THIS BOOK is to provide a comprehensive introduction to Japanese wild birds, and to those of the wider region of north-east Asia. In this book, this region is considered to include the Korean Peninsula, north-east China (the provinces of Jilin, Heilongjiang and Liaoning) and the Russian Far-east (east of 120° E) north to the Arctic coast, incorporating Ussuriland, Sakhalin, Kamchatka and Chukotka.

The *Japanese Official Bird List* (6th edition; 2000) lists 542 bird species as having occurred on the islands. In this book, I have added a further 82 species to this list, resulting in a total of approximately 620 species, of which 554 have been selected to appear in the main text. These are species that would normally be expected to be seen in Japan, plus a number of scarcer migrants and more regular vagrants; exceptional rarities are included only where there is good photographic coverage from the region. Many Japanese birds occur in other areas, especially north-east Asia. Therefore, although the main focus of the book is the Japanese avifauna, it will provide useful information for anyone planning a visit to other parts of north-east Asia. A further 28 bird species are included that one would expect to see in north-east Asia but which have not occurred in Japan, giving a total coverage of 582 species.

Currently, there is no official Rare Birds Committee in Japan. Therefore, there is no scope for review and acceptance (or otherwise) of rarity records. Some publications recognise escaped birds as native Japanese species, but in this book I have taken the approach of dealing only with species that have been officially reported, have objective evidence such as photographs, or are considered to be appropriately judged as wild birds. I hope that an official Rare Birds Committee will be established in the near future. The taxonomic sequence and nomenclature broadly follows that of *Birds of the World: A Checklist* (5th edition: 2000), better known as the Clements checklist. For common names I have tended to follow the recommendations of Gill and Wright's *Birds of the World: Recommended English Names* (2006).

Tadao Shimba

Blue-and-white Flycatcher. May, Hegura-jima

Habitats

There are in excess of 3,000 islands in the Japanese archipelago, but there are four main land masses. From north to south, these are Hokkaido, Honshu, Shikoku and Kyushu (see page 3). The distance from north to south is more than 3,000km; this vast north-south distance results in considerable climatic variation between different parts of Japan. The north-east Asian climate is generally temperate in the south and arctic in the north, with most rainfall in summer. Summers are cool and winter is very cold in the north, while summers are hot and humid and winters are cold further south. Tropical storms often hit Japan during late summer and early autumn season; these typhoons often bring pelagic species inland.

Japan itself is blessed with a range of different environments, ranging from subarctic Hokkaido to the subtropical Ryukyu Islands. Mountains account for approximately 70% of the land area and most of them are cloaked by dense woodlands.

Open seas

Japan is surrounded by nutrient-rich oceans, and these seas are excellent feeding places for seabirds, especially in the North Pacific where Black-footed and Laysan Albatrosses can be seen all-year round. Between spring and autumn Streaked Shearwater and several species of storm-petrels can be seen; these visit Japan's shores as breeding birds, while other summer visitors, such as Sooty Shearwater and Providence Petrel, breed in the southern hemisphere. In spring, pelagic species heading for their Arctic breeding grounds, such as Long-tailed and Arctic Skuas, pass through Japan, while species such as Fork-tailed Storm-petrel and Northern Fulmar visit the islands in winter.

Seabirds are best seen in Japan when travelling on ferries to Hokkaido or the Ogasawara Islands, or by ship en route to the Izu Islands.

Kazuyasu Kisaichi

Dark-morph Northern Fulmar. October, off Hokkaido

Oceanic Islands

There are many islands in Japan; those that lie some way away from the main Islands, such as the Izu, Ogasawara, Amami and Ryukyu island groups, hold significant numbers of endemic species and subspecies. The dominant biome on most of these islands is subtropical evergreen forest, a rich habitat. In addition to woodland birds, these islands hold important populations of seabirds such as Roseate Term, White-naped Tern and Brown Booby, which nest in coastal areas. The world's largest remaining population of Short-tailed Albatross breeds on Torishima, off southern Honshu.

Mitsuo Imai

Bonin Honeyeater. August, Haha-jima

Alpine areas

In central Honshu, there is a range of mountains more than 2,500m high, known as the Nihon Alps. Rock Ptarmigan live on these highlands, as do Alpine Accentor and Spotted Nutcracker. There are also alpine regions in central and eastern parts of Hokkaido, where Japanese Accentor and Pine Grosbeak breed.

Tadao Shimba

Japanese Accentor. December, Aichi

Tadao Shimba

Japanese Grosbeak. March, Aichi

Mixed forests

The northern and central parts of Honshu are covered by broad-leaved deciduous forests, coniferous forests. or a mixture of these types of woodlands. Many migrants from south-east Asia breed in these forests. Fewer species occur in the winter, but winter finches and buntings are often seen, as well as resident species such as Varied Tit, Eurasian Nuthatch, Japanese Grosbeak and Eurasian Treecreeper. Forest plantations in the foothills hold a far lower diversity of species.

Evergreen forests

South-western parts of Honshu, Shikoku and Kyushu are cloaked by evergreen forests, but species diversity is relatively low in these areas. Japanese White-eye and Varied Tit are residents, while migrant species such as Japanese Paradise Flycatcher and Fairy Pitta visit in the summer to breed.

Yoshihito Goto

Japanese White-eye. February, Amami-Oshima

Kazuyasu Kisaichi

Spectacled Guillemot. July, Hokkaido

Coastal waters and coastal areas

Gulls and cormorants can be seen in coastal waters all year round, while many species of ducks overwinter in these areas. Steller's Sea-eagle and White-tailed Eagle visit the shores of Hokkaido to winter. Japanese Cormorant, Spectacled Guillemot and Tufted Puffin breed on rocky coasts in Northern Japan, but several species of coastal birds such as auks have declined markedly. Coastal areas in southern Japan are breeding strongholds for species such as Kentish Plover, Blue Rock Thrush and Pacific Reef Heron; Kentish Plover, which breeds near the shoreline, has also declined dramatically in recent years.

Tidal flats

During Japan's period of rapid industrial growth in the latter half of the 20th century, many tidal flats were reclaimed for industrial use. Approximately 40% of the total area of tidal mudflat has disappeared in just 50 years. The remaining areas provide important feeding grounds for herons and egrets, and for the many species of waders that pass through in spring and autumn. These flats provide wintering grounds for ducks, as well as waders such as Grey Plover, Dunlin and Kentish Plover, and other species such as Saunders's Gull.

Manabu Totsuka

Dunlins. February, Aichi

Wetlands

Many wetlands in Japan have been reclaimed for agriculture, and the total area has been steadily reducing for centuries, though this rate of loss has accelerated in recent decades. Oriental Reed Warbler and Zitting Cisticola breed in reedbeds, such as those found at the edge of river terraces. Hokkaido remains relatively rich in wetland habitats; Red-crowned Cranes breed in wetlands to the east of the island. Several large wetlands remain in central and northern Honshu; Japanese Marsh Warbler, Japanese Reed Bunting and Yellow Bittern breed in these places. Lakes throughout Japan support many ducks and swans in winter, which fly south from Siberia. Swans are common winter visitors throughout northern Japan; local people often provide food for them and for the accompanying migrant ducks, allowing close encounters for photographers.

Tadao Shimba

Tundra Swans. February, Nagano.

Manabu Totsuka

Wintering cranes. Arasaki, Kagoshima

Cultivated fields

Agriculture, especially rice cultivation, takes place throughout Japan. Rice paddies play an important role for many species of birds, especially waders, rails and ducks, since they effectively act as seasonal wetlands. Paddy fields provide crucial breeding habitat for species such as Spot-billed Duck, Greater Painted-snipe, Grey-headed Lapwing and Ruddy-breasted Crake. However, in recent years farmers have tended to drain water from the rice fields as soon as farming is over for the season. The number of birds in paddy fields has declined as a result. In winter, the paddy fields provide wintering quarters for visiting passerines such as Dusky Thrush and Russet Sparrow; geese often winter in the rice paddies, especially in northern Honshu and Kyushu, and they are also important for cranes.

Tadao Shimba

Eurasian Bullfinch. March, Aichi

Urban areas

Japanese cities often have many trees and large areas of parkland, allowing people close access to birds. Chestnut-eared Bulbul, Great Tit, Oriental Turtle Dove and Japanese Pygmy Woodpecker all occur throughout the year, and in winter they are joined by species such as Dusky and Pale Thrushes, Daurian Redstart, Red-flanked Bluetail and Black-faced Bunting. Barn Swallow, a summer visitor, often nests under the eaves of houses, and Brown Hawk Owl occurs in areas with larger trees, such as around shrines and temples. Park ponds are important for wintering ducks such as Tufted Duck, Common Pochard and Northern Pintail.

Migration

Many familiar birds in Japan are migrants. The distances covered by migrants varies greatly depending on the species. Some breed in Japan and fly to south-east Asia or even further south to Australia to spend the winter in warmer climates. There are others that come to Japan from Siberia to avoid the severe northern winter conditions. Waders visit Japan in spring and autumn on the journey between Siberia and Australia or south-east Asia.

Spring migration starts at the end of March and continues until early June. Barn Swallow and Spotted Redshank are among the first to arrive in Japan in March, and the diversity of migrants increases in April. The peak time of spring migration is early May; islands in the Sea of Japan, such as Tsushima, Hegura-jima and Tobishima, are particularly rich migrant hot-spots. The majority of these spring migrants breed at higher latitudes or in northern Japan,

Tadao Shimba

Chestnut Bunting. May, Hegura-jima

Tadao Shimba

Red-necked Stint. May, Aichi

Nobuo Shiraki

Oriental Honey-buzzard. September, Nagasaki

though there are some species that can be seen around towns and cities throughout the islands, such as Barn Swallow, Eurasian Tree Sparrow, White-cheeked Starling and Oriental Turtle Dove. Autumn migration starts with waders heading south from the end of July; the peak time for this passage is early October. Autumn migration, especially around the end of September into early October, is notable for the large numbers of migrant raptors that can be seen, with species such as Grey-faced Buzzard and Oriental Honey-buzzard.

Late October sees the arrival of wintering swans, geese, ducks and gulls, while Steller's Sea Eagle and White-tailed Sea Eagle arrive in Hokkaido, and cranes fly in to Kyushu. In February, Steller's Sea Eagles and the cranes fly back to their breeding areas, but some species, such as Dusky Thrush and Daurian Redstart, stay in Japan until around April.

Teruaki Ishii

Long-tailed Duck. February, Hokkaido

Conservation

Japan has seen unprecedented industrial growth over the last 50 years. Due to the use of agricultural chemicals and habitat destruction through human activities such as deforestation and wetland reclamation, the numbers of many species of birds in Japan have declined. The once-regular visits of Swan Geese to Tokyo Bay are now very much in the past, and breeding populations of Crested Ibis and Oriental Stork, which were once commonly seen in many Japanese districts, are now extinct. However, conservation organisations in Japan recognise the decline of many species of birds, and there are efforts in place to protect a number of species at state level. As regards the Crested Ibis and Oriental Stork, captive breeding programmes are in place and there are hopes that both species can be reintroduced to the wild in Japan in the near future.

Tokio Sugiyama

Tiger Shrike. June, Nagano

Sadly, the destruction of important habitats continues. In recent years, the Isahaya Gulf was reclaimed and an important feeding and breeding ground for waders, ducks and herons disappeared. South Korean wetlands are similarly threatened, with reclamation still slated for the magnificent Saemangeum estuary despite fierce opposition. However, some development plans have been cancelled and the outlook is not all bad. Ramsar-registered sites have increased in number significantly. I hope that important places for birds will become more secure in the near future, but expanding agriculture and increasing mining activities are an ever-present threat to habitats throughout north-east Asia. Seabirds are also under threat; there have been massive declines in guillemot and puffin populations, and the expanding fishing industry may well be responsible.

Some species that breed in Japan but winter in China or south-east Asia, such as Yellow-breasted Bunting and Tiger Shrike, have suffered a population collapse in recent years. Their breeding habitats have not changed much, so presumably there must be something seriously wrong on the wintering grounds. An international effort is needed to recognise and resolve these problems. A uniquely Japanese issue is the concentration of cranes in just one small area in winter; for example, 90% of the world's Hooded Cranes and 50% of all White-naped Cranes spend the winter in Arasaki, Kyushu. The spread of an infectious disease through the birds at this wintering site would be catastrophic. Japanese conservation biologists are working hard to separate the wintering aggregations of cranes into smaller, more dispersed groups.

Where to see birds in north-east Asia

Birding can be productive almost any time of the year in Japan and South Korea if you visit the right place, but winter and spring are the times when most species are about. There are many good places to see wintering waterbirds. At Arasaki, Kyushu, thousands of cranes can be seen at close range, while enormous flocks of Baikal Teal can be watched at Geocheonnam Lake, Haenam. In spring, visiting small islands off the coast in the Sea of Japan and Yellow Sea, such as Hegura-jima and Socheong, is a great way to see migrant songbirds with a good chance of rarities.

Ran Schols

Chinese Hill Warbler. May, Hebei, China

Oceanic islands located in the south of Japan offer an opportunity to see rare endemic species; Miyake-jima is the most popular destination. It can easily be reached by overnight ferry from Tokyo (departs daily); Izu Islands Thrush and Ijima's Leaf Warbler are abundant there in spring. There is no regular organised pelagic trip available at present, but seabirds can usually be seen from the large ferries that travel between Honshu and Hokkaido or to the Ogasawara Islands.

In north-east China, Honghe in Heilongjiang province is famous for its breeding water birds, but winter birding is not productive. It is easy to get around Japan and South Korea on your own, though it is recommended to join an organised tour if visiting other areas. Spring and summer tours of Sakhalin and Ussuriland are regularly organised.

Tadao Shimba

Japanese Wood Pigeon. July, Miyako-jima

Mitsuo Imai

White-backed Woodpecker. May, Amami-Oshima

Photographing birds

As environmental awareness increased in Japan in the 1980s, more and more people began to take up birdwatching. Birding in Japan differs a little from its European counterpart in that a very high proportion of Japanese birders also take photographs – the Japanese have always been fond of photography. This means that in terms of high-quality images, the Japanese avifauna is exceptionally well-covered. Due to the technological improvement of the digital camera in recent years, we are now able to take many wonderful pictures, especially of birds in flight, and many of these appear in this book.

Photographing wild birds is not easy as they tend to fly away if approached. It is important to approach slowly and quietly as much as possible and also avoid sudden movements or noises. However, it is much easier if you know the places where birds come close to you. Feeding wild birds in winter is common in Japan, and places where this takes place usually offer good opportunities for taking photographs. Ducks can be seen at close range in many city parks, where rare species are sometimes detected among commoner birds. Swans and cranes are fed by local people in a number of places, which have sometimes become tourist attractions. Hooded and White-naped Cranes in Arasaki and Red-crowned Crane in eastern Hokkaido are among the best-known feeding spots, and these places are also good for other species that take advantage of the abundance of food. For instance, White-tailed Eagles compete with Red-crowned Cranes for fish at the Akan feeding station in Hokkaido; the eagles often stay near the feeding station throughout the winter and offer great photographic opportunities.

Nobuo Shiraki

Red-necked Phalarope. May, Aichi

Tadao Shimba

Ruddy Turnstones and a Terek Sandpiper in flight. May, Aichi

Winter finches are popular targets for photographers in Japan, and several hotels in Nagano feed these birds to offer photo opportunities. These places become very popular when winter finches are abundant, particularly Pallas's Rosefinch. Using a hide is often the best (or only) way to get good photographs at a nest or bathing point, but this is not a popular method in Japan; taking photographs at the nest is not to be encouraged since the disturbance to the adults can only endanger eggs or young. However, using a vehicle as a mobile hide is good for photographing water birds, especially on rice fields. Networks of back-roads around the rice fields allow the birder to approach waders or egrets by car. There are no serious trespassing issues as long as farming activities are not disturbed.

Teruaki Ishii

Pallas's Rosefinch. March, Kagawa

Many of the small islands in the Sea of Japan are famous for songbird migration, and they also offer great photo opportunities for a number of

Tokio Sugiyama

Oriental Stork. December, Shizuoka

Nobuo Shiraki

Brown Hawk Owl. June, Osaka

rare species. They often arrive exhausted and allow a close approach; however, regardless of the situation, it is still advisable to carry a long telephoto lens for wild bird photography in Japan.

Almost all of the pictures in this book were taken of wild birds in natural conditions. We have used photographs of birds that were captured in banding stations only when images of birds in the wild were simply unavailable. The photographs were taken in Japan or north-east Asia wherever possible; however, occasionally photographs of birds from other regions were used when images of birds in the wild in our region were not available. In these cases, different races from the ones that occur in north-east Asia may appear, though I have tried to keep this to a minimum.

Nobuo Shiraki

Long-billed Dowitcher. May, Aichi

Photographs by the author were produced using Canon EOS3 (film) and EOS 30D (digital) cameras with an EF 500mm f4 IS lens and, in most cases, with a 1.4 x extender attached. The majority (about 70%) of images in this book were taken using film (Fujichrome 100 Provia), with the rest taken digitally. With autofocus performance significantly improving over recent years, the key for success in wild bird photography is to minimise camera shake. I recommend the use of a tripod wherever possible, and to use a higher shutter speed as much as you can. Nowadays, some lenses are equipped with stabilisers which are particularly useful for difficult images of birds in flight. Flash is only used for night photography in Japan; the images can look artificial if used in daylight.

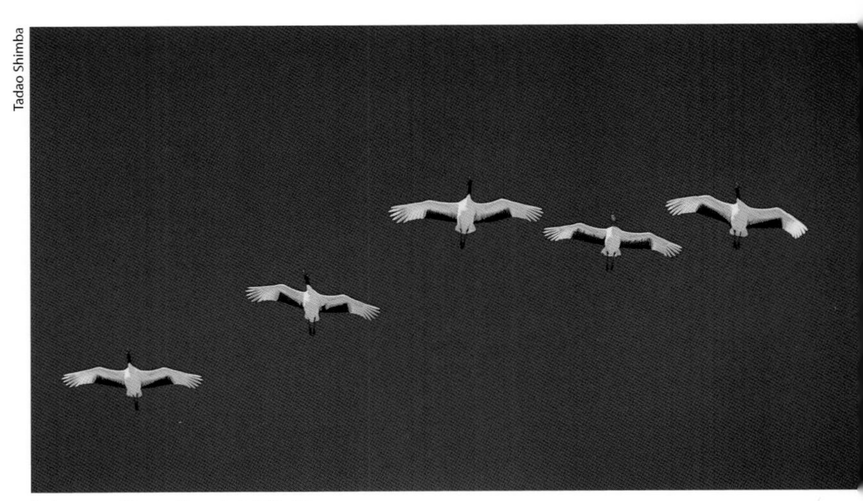

Tadao Shimba

Red-crowned Cranes. December, Hokkaido

HOW TO USE THIS BOOK

Measurements
The length of the bird is shown here; where this differs between males and females separate figures are given. Wingspan (WS) is given for groups where this measurement is of use for identification.

354 *BIRDS OF JAPAN AND NORTH-EAST ASIA*

Pale Thrush

Turdus pallidus 23–25cm

Description Male has dark-greyish head and throat, olive-brown back and dark tail with white corners. Also buffy-grey on breast and flanks with white belly and undertail-coverts. Female has olive-brown head and back, pale supercilium, white throat with brown moustachial stripe and paler underparts.
Voice Song is a loud melodious whistle and call is a harsh *zeeeh* or loud and explosive *kwak-kwak-kwak* when alarmed.
Similar species Brown-headed Thrush has rufous flanks and no white on tail.

Range Breeds from the Amur basin east to Ussuriland, south to north-eastern China and Korea. Winters in southern Korea, Japan, Taiwan and south-eastern China.
Status in Japan Common winter visitor to wooded areas from central Honshu southwards. Common in city parks and wooded areas. A few breed in Hiroshima.

Takeo Sugiyama
Adult ♂. January, Kyoto

Tadao Shimba
Adult ♀. February, Aichi

Yoshihito Goto
Adult ♂. February, Amami-Oshima

Description
gives identification features of use in the field, with separate sections on voice and confusion species.

Green box text
contains information on the world distribution of the species, plus the bird's status in Japan.

Localities in Japan
are given by prefecture; see the map on page 3. Localities in other parts of the region are followed by the country name.

MAPS

Yellow shading
shows the range of summer migrants that usually visit the region for breeding.

Blue shading shows areas where the species occurs in winter.

Green shading
indicates a resident population, or regular occurence throughout the year, though numbers present may vary.

Island populations
are not generally mapped in detail; a mark on the island indicates a bird's status.

All species maps were created by Takaaki Inoue and Takefumi Abe (BIONOMICSLAB Co. Ltd)

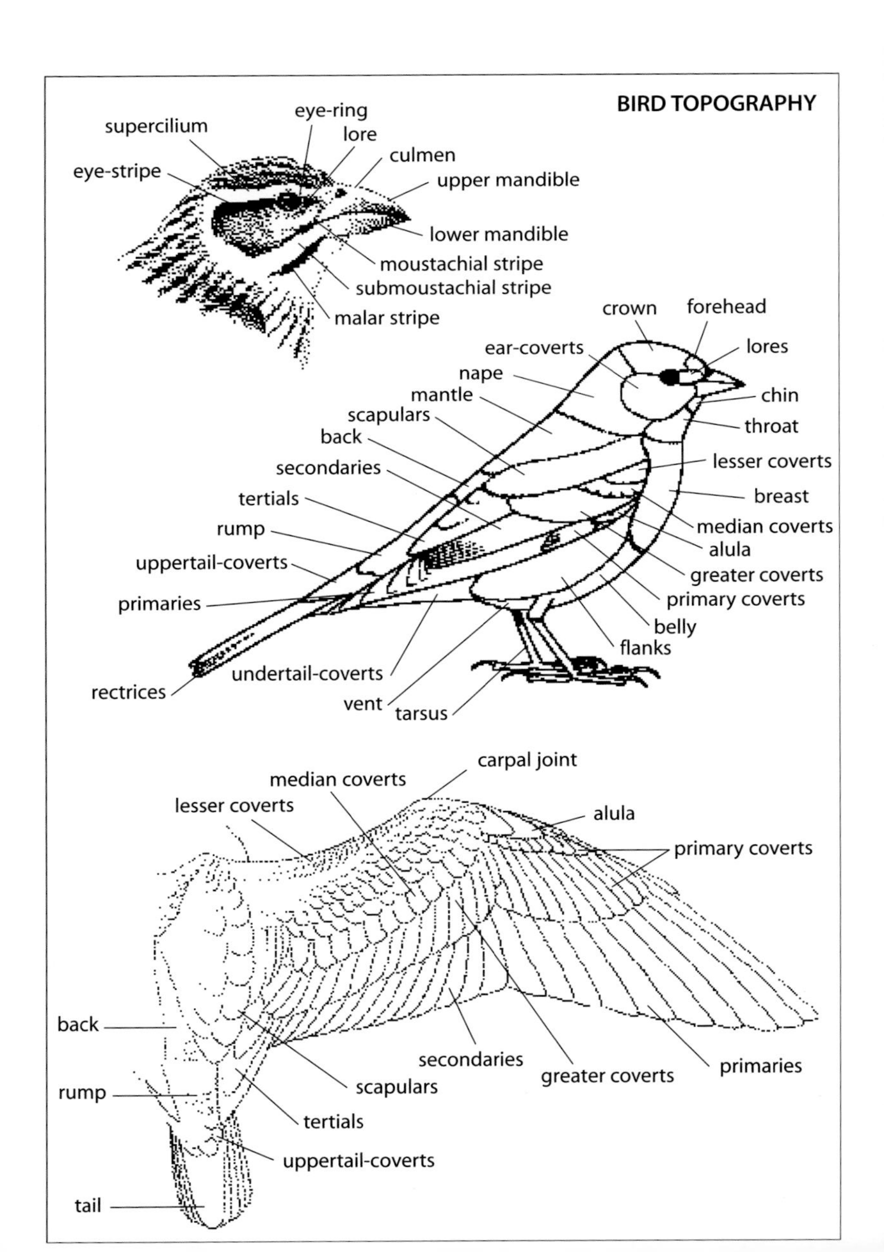
BIRD TOPOGRAPHY
supercilium
eye-ring
lore
culmen
eye-stripe
upper mandible
lower mandible
moustachial stripe
submoustachial stripe
malar stripe
crown
forehead
lores
ear-coverts
nape
chin
mantle
scapulars
throat
back
secondaries
lesser coverts
breast
tertials
median coverts
rump
alula
uppertail-coverts
greater coverts
primaries
primary coverts
belly
flanks
undertail-coverts
rectrices
vent
tarsus
carpal joint
median coverts
lesser coverts
alula
primary coverts
back
secondaries
greater coverts
primaries
scapulars
rump
tertials
uppertail-coverts
tail

Tadao Shimba

Mugimaki Flycatcher. May, Hegura-jima

Red-throated Diver

Gavia stellata 63cm

Description Small diver with slender neck and uptilted dark bill. In breeding plumage has plain grey head with distinctive reddish-brown throat patch. In non-breeding plumage shows dark greyish-brown head, hindneck and back contrast with white face and foreneck. Juvenile has greyish-brown head and greyish-brown back finely marked with white.

Similar species Both Black-throated Diver and Pacific Diver have larger heads, darker necks and straight bills.

Akira Yamamoto

Adult non-breeding. January, Osaka

Range Breeds across northern Eurasia, northern North America and Greenland. Winters on large lakes and in coastal waters in temperate areas.

Status in Japan Uncommon winter visitor to coastal waters from Kyushu northwards.

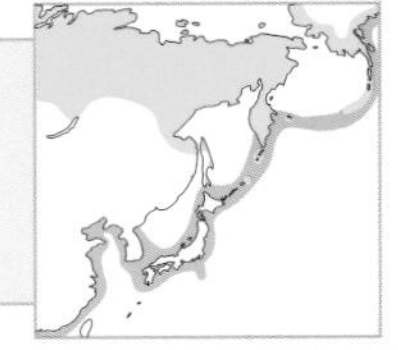

Tokio Sugiyama

Adult non-breeding. January, Osaka

Black-throated Diver

Gavia arctica 72cm

Tadao Shimba

Adult breeding. May, Aichi

Description Medium-large diver with straight, dark bill. In breeding plumage has plain grey head, dark foreneck with glossy-green patch, and fine dark streaks on sides of neck. In non-breeding plumage shows dark head, hindneck and back contrast with white foreneck and underparts. Juvenile paler with buff-coloured edges to scapulars. White flanks conspicuous when swimming.

Similar species Pacific Diver is slightly smaller, has a more rounded and paler head, shorter and finer bill and dark flanks; often acquires dark throat strap in non-breeding plumage.

Akira Yamamoto

Adult breeding. May, Aichi

Range Breeds across northern Eurasia east to far-eastern Russia, south to Mongolia, lower Amur Basin and Sakhalin. Winters on large lakes and in coastal waters in temperate areas.

Status in Japan Common in winter from coast of Kyushu northwards. May occur on fresh-water lakes.

Mitsuo Imai

Adult non-breeding. May, Tsushima

Pacific Diver

Gavia pacifica 65cm

Description Medium-sized diver with straight, dark bill. In breeding plumage has plain grey head (paler than Black-throated) and dark foreneck with glossy-purple patch, with fine dark streaks on both sides of neck. In non-breeding plumage dark head, hindneck and back contrast with white foreneck and underparts; often acquires dark throat strap on foreneck. Juvenile browner.
Similar species Black-throated Diver is slightly larger, has a more peaked forehead, larger bill and conspicuous white flanks.

Kazuyasu Kisaichi

Adult breeding. February, Ibaragi

Range Breeds in northern North America and north-eastern Russia. On both coasts of the Pacific in winter.
Status in Japan The most common diver in Japan. Common winter visitor to coastal waters from Kyushu northwards. Occasionally occurs on freshwater lakes.

Kazuyasu Kisaichi

Adult non-breeding. April, Ibaragi

White-billed Diver

Gavia adamsii 89cm

Mitsuo Imai

Adult breeding. May, Ibaragi

Tokio Sugiyama

Adult breeding. May, Ibaragi

Description Large diver with massive yellow upcurved bill. In breeding plumage has glossy-black head and neck with white necklace and two striped half-collars, back checkered with black and white. In non-breeding plumage shows dark greyish-brown head, hindneck and back contrast with white foreneck and underparts. Juvenile has pale edges to back feathers and ivory coloured bill.

Similar species Both Black-throated and Pacific Divers are smaller and have dark, straight bills. Great Northern Diver *G. immer*, recently reported from northern Japan, has straight greyish bill.

Range Breeds in arctic Siberia and North America. Winters on northern waters along the Pacific and Atlantic coasts.

Status in Japan Uncommon winter visitor to coastal waters in northern Japan.

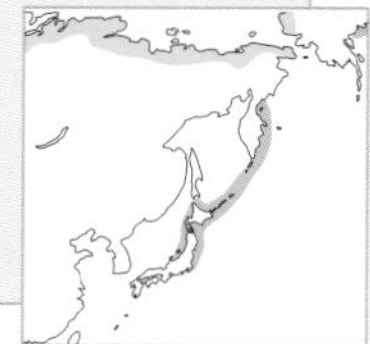

Shigeo Ozawa

First-winter. February, Chiba

Little Grebe

Tachybaptus ruficollis 26cm

Description Small grebe with short neck and bill. Breeding plumage is dark brown overall, with chestnut cheeks and neck, and yellow gape patch at the base of bill. Cheeks and foreneck are paler in non-breeding plumage. Shows no white on wings in flight.
Similar species Both Slavonian and Black-necked Grebes are larger, have longer necks and darker backs.

Tokio Sugiyama

Adult breeding. July, Aichi

Range Widespread in temperate and tropical Eurasia from Europe east to China and Korea, south to southern Asia and Africa. Partly migratory.
Status in Japan Common resident on well-vegetated ponds and freshwater marshes. Often breeds on ponds in city parks. In winter, appears on more open waters, often in flocks on lakes and estuaries.

Akira Yamamoto

Adult breeding. June, Aichi

Tokio Sugiyama

First-winter. February, Aichi

Red-necked Grebe

Podiceps grisegena 47cm

Manabu Totsuka

Adult breeding. June, Hokkaido

Description Medium-large grebe with long neck and straight, dark-tipped bill with yellow base. In breeding plumage, dark head and greyish-white cheeks contrast with chestnut neck and breast, yellow base to bill and dark-brown back. In non-breeding plumage, dark greyish-brown neck, white cheeks and dark-tipped yellow bill.
Similar species Great Crested Grebe is larger, has longer neck, and whiter face and foreneck.

Shigeo Ozawa

Adult non-breeding. February, Ibaragi

Range Breeds from Europe to western Siberia, then from eastern Siberia to the Okhotsk coast and Kamchatka Peninsula, south to Amur Basin, north-eastern China and Sakhalin; also in northern North America. Winters south to the temperate zone.
Status in Japan Breeds on lakes and marshes in Hokkaido; also common winter visitor to coastal waters from Kyushu northwards.

Akira Yamamoto

Adult breeding. April, Aichi

Great Crested Grebe

Podiceps cristatus 56cm

Description Large grebe with long neck and straight, pinkish bill. In breeding plumage has black-edged chestnut tufts around face with black crest. In non-breeding plumage, black crown and white face with dark lores, dark-brown hindneck and back contrast with white foreneck.
Similar species Red-necked Grebe is smaller with shorter neck and greyer foreneck.

Akira Yamamoto

Adult breeding. May, Aichi

Range Breeds across Eurasia from Europe east through Siberia to Okhotsk coast, south to central Asia, Mongolia, northern China and Ussuriland; also in Australia and Africa. Winters on large lakes and in coastal waters in temperate zone.
Status in Japan Common winter visitor to coastal waters, rivers and lakes from Kyushu northwards. Small numbers breed on well-vegetated lakes in Aomori and Shiga.

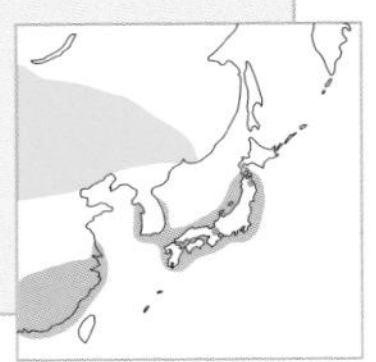

Teruaki Ishii

Adult non-breeding. March, Shiga

Akira Yamamoto

Adult breeding. May, Aichi

Black-necked Grebe

Podiceps nigricollis 31cm

Kazuyasu Kisaichi

Adult breeding. March, Ibaragi

Description Small grebe with slightly upcurved, pointed black bill and red eye. In breeding plumage black above and chestnut below, with golden ear-tufts. In non-breeding plumage, dark-grey head, ear-coverts, hindneck and back contrast with greyish-white chin and fore-neck. In flight white patch on secondaries is visible.
Similar species Non-breeding Slavonian Grebe has whiter cheeks and foreneck; also straighter bill.

Range Breeds from Europe east across Siberia to about Ob River, then in north-eastern China and Ussuriland; also in western North America and southern Africa. Winters south to temperate areas.
Status in Japan Common winter visitor to coastal waters, estuaries and freshwater lakes. Large flocks often seen in spring.

Akira Yamamoto

Adult non-breeding. December, Aichi

Slavonian Grebe

Podiceps auritus 33cm

Other name Horned Grebe
Description Small grebe with red eyes and pale tipped, straight dark bill. In breeding plumage has dark head with golden-yellow ear-tufts and chestnut neck. In non-breeding plumage, dark cap and hindneck contrast with white cheeks and foreneck. In flight, white patch on secondaries is visible.
Similar species Non-breeding Black-necked Grebe has darker cheeks, greyer foreneck and slightly upcurved bill.

Tadao Shimba

Adult breeding. April, Aichi

Kazuyasu Kisaichi

Adult non-breeding. April, Chiba

Range Breeds from northern Europe east through Siberia to the Okhotsk coast and Kamchatka Peninsula, south to Lake Baikal, Amur Basin and northern Sakhalin; also in northern North America. Winters south to the temperate zone.
Status in Japan Uncommon winter visitor to coastal waters, rarely occurs on inland waters.

Wandering Albatross

Diomedea exulans 120cm WS300cm

Description Very large seabird. Adult has almost entirely white body and wings with dark trailing edges to secondaries, black primaries and pale pinkish bill. Takes many years to acquire adult plumage. Immature has white face, dark-brown back and body, dark-brown upperwings, white underwings with dark trailing edges to secondaries and black primaries.
Similar species Immature Short-tailed Albatross is smaller and has mottled white and dark brown underwings.

Tadao Shimba

Adult. December, off Sydney, Australia

Range Typically occurs in southern hemisphere north to about the Tropic of Capricorn.
Status in Japan One record of two immatures captured in 1970 in East China Sea.

Short-tailed Albatross

Diomedea albatrus 91.5cm WS210cm

Shigeo Ozawa

Adult. April, Torishima

Description Largest breeding seabird in the northern hemisphere. Adult has mostly white body with large pinkish bill, yellow head and hindneck, dark outer wings that contrast with white inner half, white underwings with black margins, black tail and blackish-grey legs. First–year birds are dark-brown overall with pink bill, pale belly and white flashes on upperwings. Adult plumage is attained after several years; amount of white on wings gradually increases with age.

Similar species Black-footed Albatross is smaller, has darker bill and legs, and uniform dark-brown wings with white flash at the base of primaries.

Kazuyasu Kisaichi

Adult. April, Torishima

Range Breeds on oceanic islands in Pacific Ocean and East China Sea then disperses to northern Pacific.

Status in Japan Mainly breeds on Torishima. Recovering from brink of extinction, world population an estimated 1,000 birds. Rarely seen in pelagic waters off Pacific coasts in spring.

Kazuyasu Kisaichi

Immature. April, Torishima

Laysan Albatross

Diomedea immutabilis 80cm WS200cm

Description Upperwings uniform dark-brown with white flash at base of primaries. White head and body with pinkish bill, dark eye-patch, pinkish feet and dark tail. White underwings with black margins and black patches.
Similar species Immature Short-tailed Albatross is larger, has pink bill and white patches on upperwings.

Kazuyasu Kisaichi

Adult. June, Hokkaido

Range Widely distributed across northern Pacific Ocean.
Status in Japan Small numbers breed on oceanic islands in winter and can be seen year-round in pelagic waters, mainly off Pacific coasts.

Kazuyasu Kisaichi

Adult. June, Hokkaido

Kazuyasu Kisaichi

Adult. June, Hokkaido

Black-footed Albatross

Phoebastria nigripes 78.5cm WS180cm

Kazuyasu Kisaichi

Adult. October, off Hokkaido

Description Large dark-brown seabird with black bill and white patches at the base of bill, white spot just below eye, white uppertail-coverts and white outer primary shafts above and below. Immature is similar to adult, but has paler face and darker uppertail-coverts.
Similar species Immature Short-tailed Albatross is larger, has pink bill and feet and white patches on upperwings.

Manabu Totsuka

Adult. April, Hawaii, USA

Range Widely distributed across northern Pacific Ocean.
Status in Japan Breeds on oceanic islands in winter and can be seen year-round in pelagic waters, mainly off Pacific coasts.

Kazuyasu Kisaichi

Adult. July, off Hokkaido

Northern Fulmar

Fulmarus glacialis 49.5cm WS107cm

Description Stocky seabird with short broad wings, stout yellowish bill with prominent nasal tube and yellowish legs. Flight distinctive; long glides with occasional stiff, shallow wingbeats. Two distinct colour morphs; dark morph is dark-brown overall with pale patches at the base of primaries; light morph has white head, grey upperwings and dark tail.
Similar species Dark shearwaters have longer and narrower wings.

Kazuyasu Kisaichi

Pale morph. June, off Hokkaido

Range Widely distributed across northern Pacific Ocean.
Status in Japan Often seen in pelagic waters off the Pacific coasts of northern Japan; dark-morph birds are more commonly seen.

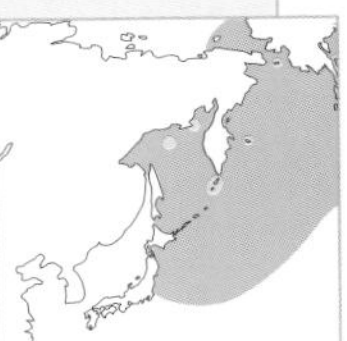

Kazuyasu Kisaichi

Dark morph. October, off Hokkaido

Mottled Petrel

Pterodroma inexpectata 35cm WS85cm

Description Medium-sized grey petrel with diagnostic dark-grey belly patch, and white underwings with distinctive black bar. Dark line across upperwings forms an M-shaped mark.

Range Breeds on oceanic islands off New Zealand then disperses north across Pacific Ocean after breeding.
Status in Japan One specimen collected after typhoon in Hiroshima in 1986; few other sightings.

Brent Stephenson

Adult. February, off New Zealand

Providence Petrel

Pterodroma solandri 40cm WS100cm

Tadao Shimba

Adult. October, off Sydney, Australia

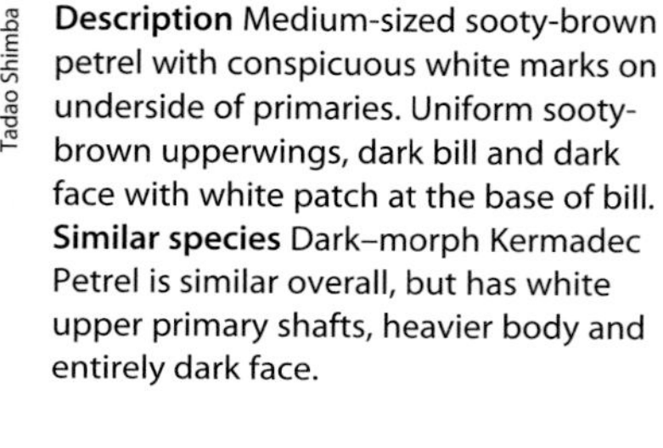

Description Medium-sized sooty-brown petrel with conspicuous white marks on underside of primaries. Uniform sooty-brown upperwings, dark bill and dark face with white patch at the base of bill. **Similar species** Dark–morph Kermadec Petrel is similar overall, but has white upper primary shafts, heavier body and entirely dark face.

Tadao Shimba

Adult. October, off Sydney, Australia

Range Breeds on Lord Howe Island in south-western Pacific then disperses north across Pacific Ocean after breeding.
Status in Japan Rare summer visitor to pelagic waters off east coast of northern Japan; mostly recorded in August and September.

Black-winged Petrel

Pterodroma nigripennis 31cm WS66cm

Tadao Shimba

Adult. January, Lord Howe, Australia

Description Medium-small petrel with grey upperparts and white underparts. Black line across upperwings forms clear M-shaped pattern in flight. Also has distinctive black bar on underwings. Grey head with black eye-patch and dark bill. **Similar species** Bonin Petrel has darker crown, neck-sides and upperwings.

Range Breeds on oceanic islands in southern Pacific then disperses north across Pacific Ocean after breeding.
Status in Japan One specimen collected after typhoon in Hokkaido in 1980; few other sightings.

Kermadec Petrel

Pterodroma neglecta 38cm WS90cm

Description Medium-sized dark-brown petrel with white primary shafts above, and white marks on underside of primaries, dark bill, dark face and heavy body. Pale morph has white underparts.
Similar species Providence Petrel has slimmer body and lacks white primary shafts above.

Tadao Shimba

Adult, dark morph. January, Lord Howe, Australia

Range Breeds on oceanic islands in southern Pacific then disperses north across Pacific Ocean after breeding.
Status in Japan Very rare summer visitor to pelagic waters off east coast. Dark morph is most often seen.

Tadao Shimba

Adult intermediate. January, Lord Howe, Australia

White-necked Petrel

Pterodroma cervicalis 43cm WS97cm

Description Medium-sized petrel with grey back and white underparts. Dark bill, dark cap and white neck. Dark line across upperwing forms M; faint dark bar on underwings. Frequently seen gliding and seldom flies high above waves.
Similar species Bonin Petrel is smaller, has dark neck and distinctive black bar on underwings.

Range Breeds on oceanic islands in southern Pacific then disperses north across Pacific.
Status in Japan Rare summer visitor to pelagic waters off east coast. Often observed from ferries to the Ogasawara Islands in summer.

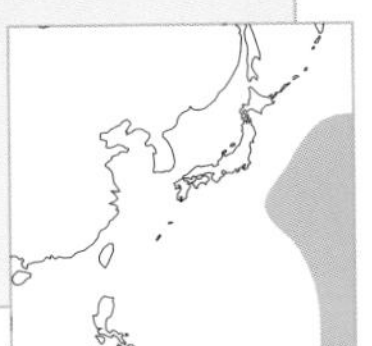

Brent Stephenson

Adult. February, off New Zealand

Bonin Petrel

Pterodroma hypoleuca 31cm WS69cm

Kazuyasu Kisaichi

Adult. April, Hawaii, USA

Mitsuo Imai

Adult. May, Ogasawara Islands

Kazuyasu Kisaichi

Adult. October, Minami-Iwojima

Description Medium-small petrel with dark upperparts, white underparts, dark bill, white forehead and dark crown, face and neck-sides. Upperwing greyish-brown with indistinct dark M-shaped pattern. Distinctive black bar on underwing. Usually seen gliding low over water.

Similar species Stejneger's Petrel *P. longirostris* (rare visitor to Japan) has clear M-shaped mark on upperwings and lacks underwing black bar. Also has faster wingbeats, and arcs high over waves. See also White-necked and Black-winged Petrels.

Range Breeds on oceanic islands in tropical Pacific then disperses to central and western Pacific Ocean.

Status in Japan Uncommon summer visitor to deep pelagic waters off east coast. Breeds on Ogasawara Islands and commonly observed in open waters around the islands.

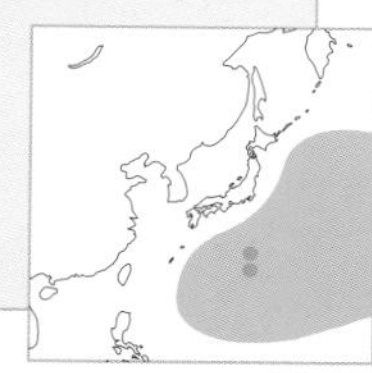

Streaked Shearwater

Calonectris leucomelas 48cm WS122cm

Description Large shearwater with brown upperparts and white underparts. Pale hooked bill, white forehead and streaked head. Dark line across upper mantle forms M-shaped pattern; also dark marks on underside of primary coverts. In flight, slow wingbeats are interspersed with gliding; arcs high above waves during rough seas.
Similar species Pale–morph Wedge-tailed Shearwater is uniform dark-brown above, has dark head and dark bill with clear black margins on underwings.

Mitsuo Imai

Adult. August, Izu Islands

Range Breeds on islands off China, Korea and Japan. Winters south to northern Australia.
Status in Japan Abundant summer visitor that breeds on small islands off coast. Most common shearwater in Japan and frequently seen from shore.

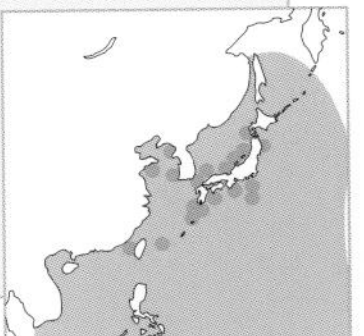

Mitsuo Imai

Adult. August, Izu Islands

Mitsuo Imai

Adult. August, Izu Islands

Flesh-footed Shearwater

Puffinus carneipes 51cm WS109cm

Tadao Shimba

Adult. December, off Sydney, Australia

Description Large dark-brown shearwater with dark-tipped pinkish bill and distinctive pink feet. Flight is buoyant with slow wingbeats.
Similar species Sooty Shearwater has dark bill and silvery white underwings. Flies with rapid wingbeats interspersed with long glides.

Tadao Shimba

Adult. December, off Sydney, Australia

Range Breeds on oceanic islands off Australia then disperses north after breeding.
Status in Japan Uncommon spring-summer visitor to pelagic waters. Seldom observed from coast.

Tadao Shimba

Adult. December, off Sydney, Australia

Wedge-tailed Shearwater

Puffinus pacificus 39cm WS97cm

Description Large shearwater with dark-brown upperparts. Pale morph has white underparts, dark head and bill, pink feet and black margins to underwings. Dark morph all dark brown. Flies with slow wingbeats interspersed with gliding. Arcs high above waves during rough seas.
Similar species Streaked Shearwater has white forehead, streaked head and dark marks on underside of primary coverts.

Mitsuo Imai

Adult, pale morph. May, Ogasawara Islands

Range Tropical and subtropical waters of Pacific and Indian Oceans.
Status in Japan Common summer visitor to breeding colonies on the Ogasawara and Iwo Islands. Pale morph is most often seen.

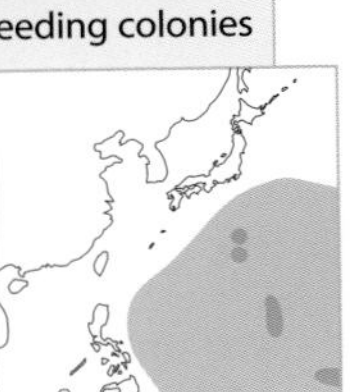

Norio Yamagata

Adult, pale morph. June, Ogasawara Islands

Norio Yamagata

Adult, pale morph. June, Ogasawara Islands

Buller's Shearwater

Puffinus bulleri 42cm WS102cm

Tadao Shimba

Adult. January, off New Zealand

Description Elegant shearwater with distinct dark M-shaped pattern on grey upperwings. Dark cap, dark bill, long grey tail and white underparts. Underwings with black margins. In flight, slow wingbeats are interspersed with gliding.
Similar species Streaked Shearwater is darker above, has white forehead and streaked head, and dark marks on under primary-coverts.

Tadao Shimba

Adult. January, off New Zealand

Range Breeds on oceanic islands off New Zealand then disperses north across Pacific Ocean after breeding.
Status in Japan Considered vagrant until 1980s but now discovered to be a regular if uncommon summer-autumn migrant off northern Pacific coasts.

Kazuyasu Kisaichi

Adult. June, Chiba

Sooty Shearwater

Puffinus griseus 43cm WS109cm

Description Large dark shearwater with pale-silvery underwings, dark bill and dark feet. In flight, fast wingbeats are interspersed with long glides.
Similar species Short-tailed Shearwater similar overall, but is smaller with more rapid wingbeats, and has shorter bill, wings and tail. Flesh-footed Shearwater has pinkish bill and feet, and longer wings.

Kazuyasu Kisaichi

Adult. July, off Hokkaido

Range Breeds on oceanic islands in southern hemisphere then disperses north after breeding.
Status in Japan Common spring-summer visitor to pelagic waters. Seldom observed from shore though flocks of thousands sometimes observed off the coast of eastern Hokkaido

Mitsuo Imai

Adult. July, Chiba

Tadao Shimba

Adult. January, off New Zealand

Short-tailed Shearwater

Puffinus tenuirostris 33cm WS97cm

Tadao Shimba

Adult. January, off Victoria, Australia

Description Medium-sized dark shearwater with pale-silvery underwings, short wings and tail, dark bill and feet. In flight, has rapid wingbeats and feet extend beyond tip of tail.

Similar species Sooty Shearwater is larger and has longer bill, tail and wings.

Tadao Shimba

Adult. January, off Victoria, Australia

Range Breeds on oceanic islands off south-eastern Australia then disperses north across Pacific Ocean after breeding.

Status in Japan Common spring-summer visitor offshore. Often abundant in late spring and flocks can sometimes be observed from the coast. Heavy mortality of juvenile birds occurs in some years.

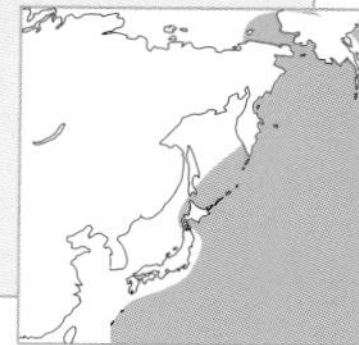

Kazuyasu Kisaichi

Juvenile. May, Chiba

Bulwer's Petrel

Bulweria bulwerii 27cm WS61cm

Description Small dark-brown petrel with long wedge-shaped tail, dark bill, brown panel across upperwing-coverts and light-brown underwings. Flies in buoyant fashion, gliding over waves.
Similar species Tristram's Storm-petrel is about the same size, but has forked tail (often difficult to detect at range) and whiter panel across upperwing-coverts.

Tadao Shimba

Adult. August, Ogasawara Islands

Range Tropical and subtropical waters of Pacific and Atlantic Oceans.
Status in Japan Common summer visitor to warm waters. Breeds on Ogasawara Islands and Iwo-jima. Often seen in pelagic waters off east coast of Japan and around Ryukyu Islands.

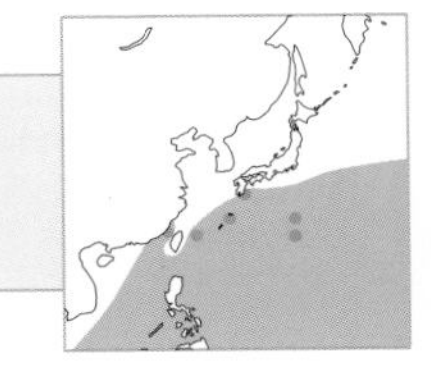

Audubon's Shearwater

Puffinus lherminieri 31cm WS69cm

Description Small shearwater with short wings, uniform brown upperparts and white underparts, brown head and dark slender bill. Underwings have white linings with dark margins. Flies low over the water with fast wing beats.
Similar species Vagrant Manx Shearwater *P. puffinus* (one record in Japan) is slightly larger with darker upperparts.

Kazuyasu Kisaichi

Adult. September, Minami-Iwojima

Range Widely distributed in tropical seas.
Status in Japan Poorly known. Uncommon at sea around Ogasawara Islands and Iwo-jima, presumed to breed on Iwo-jima. Fewer sightings in recent years.

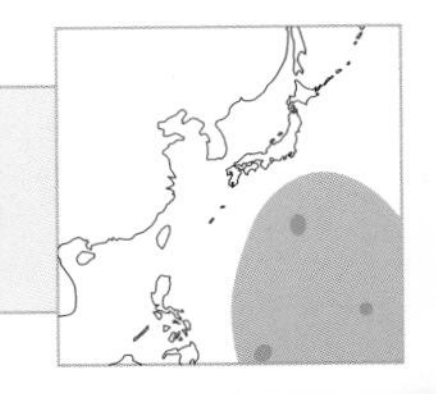

Wilson's Storm-petrel

Oceanites oceanicus 18cm WS41cm

Brent Stephenson

Adult. January, off New Zealand

Description Small dark storm-petrel with white rump, paler crescent across upperwing and square tail. In flight, glides and banks low over water and sometimes flutters over sea surface; long legs extend well beyond tail. Yellow webs of feet only visible at close range.

Similar species Other small dark storm-petrels in the region have forked tails and shorter legs.

Range Widely distributed throughout southern hemisphere and one of the most common storm-petrels worldwide.

Status in Japan Only a few records from Japan, but possibly overlooked among other small, dark storm-petrels.

Madeiran Storm-petrel

Oceanodroma castro 19cm WS46cm

Yoshio Osawa

Adult. July, Iwate

Description Medium-sized dark storm-petrel with square tail, white rump and white undertail-coverts. Brown panel across upperwing-coverts visible in flight. Flight steady and rhythmic, low over water.

Similar species Leach's Storm-petrel has forked tail, dark centre line on white rump and dark undertail-coverts.

Range Tropical and subtropical Pacific and Atlantic Oceans; north to Japan.

Status in Japan Common summer visitor to pelagic waters off northern coasts. Breeds on small islands off the Pacific coast of Honshu.

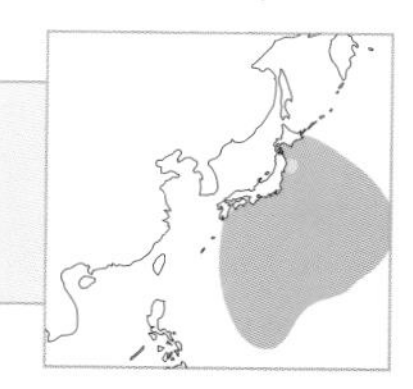

Leach's Storm-petrel

Oceanodroma leucorhoa 20.5cm WS48cm

Description Medium-sized dark storm-petrel with forked tail, white rump with dark central line, and brown panel across upperwing-coverts. Flight erratic and tern-like, low over water.
Similar species Madeiran Storm-petrel is darker, lacks central line on white rump, has square tail and normally has white undertail-coverts.

Norio Yamagata

Adult. June, off Hokkaido

Range Northern Pacific and Atlantic Oceans.
Status in Japan Common summer visitor to pelagic waters off northern coasts. Breeds on small islands off the Pacific coasts of Hokkaido and northern Honshu.

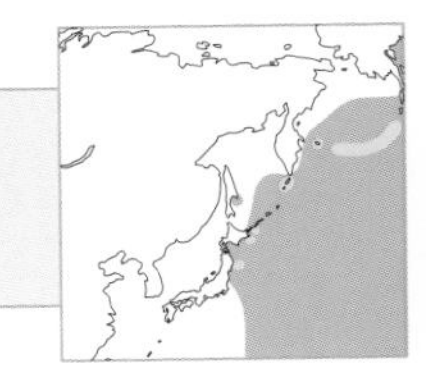

Swinhoe's Storm-petrel

Oceanodroma monorhis 19cm WS38.5cm

Description Medium-sized dark storm-petrel with forked tail and brown panel across upperwing-coverts. Indistinct white flash at the base of upper primaries is only visible at close range. In flight, glides and banks low over water and sometimes flutters over sea surface.
Similar species Tristram's Storm-petrel is larger with longer wings and tail. Separated from other small and medium-sized storm-petrels by lack of white rump.

Yoshio Osawa

Adult. July, Iwate

Range Breeds on islands off China, Korea and Japan. Winters south to Indian Ocean and Coral Sea.
Status in Japan Uncommon summer visitor to pelagic waters off northern coasts. Breeds on small islands off Honshu and Kyushu.

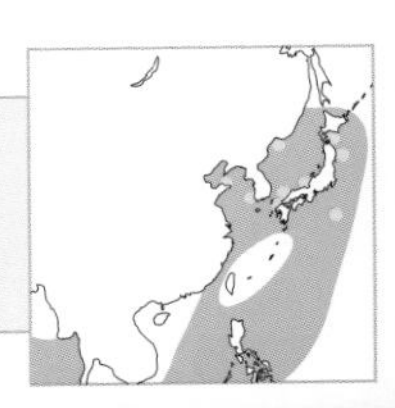

Tristram's Storm-petrel

Oceanodroma tristrami 25.5cm WS56cm

Tony Palliser

Adult. October, off Sydney, Australia

Description Large dark storm-petrel with forked tail. In flight, shows broad cream-coloured panel across upperwing-coverts. Glides and banks low over water and flies high over waves during rough seas.
Similar species Matsudaira's Storm-petrel has white flash at the base of upper primaries. Bulwer's Petrel has brown panel across upper primaries and long wedge-shaped tail.

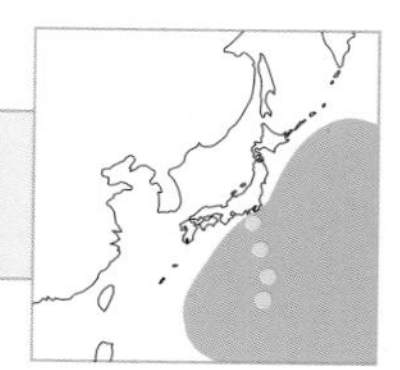

Range Pelagic in the north-western Pacific north to Japanese waters.
Status in Japan Uncommon summer visitor to pelagic waters, mainly off the Pacific coast. Often seen around Izu Islands where it breeds.

Matsudaira's Storm-petrel

Oceanodroma matsudairae 22.5cm WS56cm

Norio Yamagata

Adult. April, Ogasawara Islands

Description Large dark storm-petrel with forked tail. In flight, broad brown panel across upperwing-coverts and distinctive white flash at the base of upper primaries are visible. Flight often slow and deliberate; flies high over waves during rough seas.
Similar species Can be separated from both Tristram's Storm-petrel and Bulwer's Petrel by distinctive white flash at base of upper primaries.

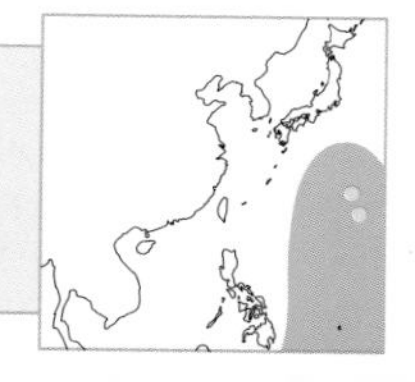

Range Breeds on oceanic islands in southern Japan. Winters south to Indian Ocean and Coral Sea.
Status in Japan Breeds on Ogasawara and on Iwo Islands in winter and can be reliably seen around those islands; rare elsewhere. Occasionally follows ships.

Fork-tailed Storm-petrel

Oceanodroma furcata 20cm WS46cm

Description Only grey storm-petrel found in Japan waters. White throat, underwings and undertail-coverts. Forked tail. In flight, glides and banks low over water.
Similar species Can be separated from other storm-petrels by grey plumage.

Kazuyasu Kisaichi

Adult. January, Ibaragi

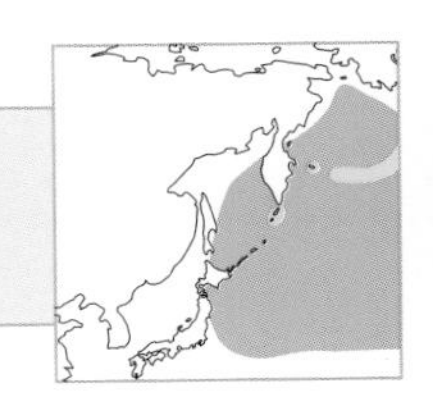

Range Typically a species of the northern Pacific Ocean and Bering Sea. Breeds on oceanic islands including Kuril Islands.
Status in Japan Uncommon winter visitor to pelagic waters off northern coasts. Occasionally seeks refuge in fishing ports during rough weather.

Masked Booby

Sula dactylatra 73cm WS145cm

Description Largest booby. White body with black tail, primaries and secondaries. Black face, yellowish-orange bill and greyish-blue legs. Juvenile has brown head and neck, white collar and underparts with mottled brown upperparts and dark tail.
Similar species Red-footed Booby has white face, red bill and white tail.

Kazuyasu Kisaichi

Immature. September, Mikura-jima

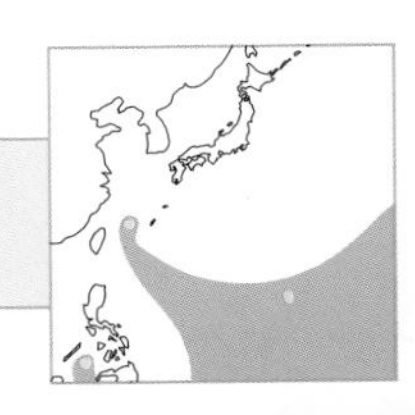

Range Tropical and subtropical Pacific, Indian and Atlantic Oceans.
Status in Japan Very rare visitor to southern pelagic waters. Few mainland records.

Red-footed Booby

Sula sula 70cm WS152cm

Mitsuo Imai

Adult. August, Ogasawa Islands

Description White morph is a large white seabird with black primaries and secondaries and white tail. Brown morph is all greyish-brown. Blue bill, red legs and feet. Juvenile is browner overall.
Similar species Masked Booby is larger, has distinctive black face, yellowish-orange bill and black tail. Juvenile Masked Booby has white collar and contrasting paler back and forewings above.

Range Tropical and subtropical Pacific, Indian and Atlantic Oceans.
Status in Japan Rare summer visitor to southern pelagic waters. Small numbers breed on Nakanougan-jima in Ryukyu Islands.

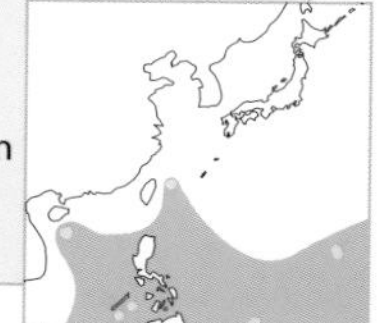

Brown Booby

Sula leucogaster 73cm WS145cm

Akira Yamamoto

Adult. June, Rota Island

Description Large seabird with long pointed bill, long wings and wedge-shaped tail. In flight, distinctive chocolate-brown head, neck and upperparts contrast with white underparts. Bill and feet yellowish-green. Juvenile has similar colour pattern to adult but has mottled underparts.
Similar species Separated from other boobies by entirely chocolate-brown upperparts.

Kazuyasu Kisaichi

Immature. September, Ogasawara Islands

Range Widely distributed in tropical and subtropical Pacific, Indian and Atlantic Oceans.
Status in Japan Locally common resident around Ogasawara Islands, Iwo Islands and Ryukyu Islands; rare elsewhere.

Dalmatian Pelican

Pelecanus crispus 160–180cm WS250cm

Description Very large greyish-white waterbird with huge, dark-grey bill and orange-yellow bill-pouch. Dark bare skin around eyes and lores, short greyish legs and short white tail. In flight, white forewings contrast with dark primaries on upperwings, while underwings are white with greyish primaries. Immature has dark brown back with pale yellow bill-pouch.
Similar species Great White Pelican is slightly smaller, has pink legs and underwings show black primaries and secondaries.

Kazuyasu Kisaichi

Probable first-winter. January, Ibaragi

Range Breeds in temperate Eurasia east to Mongolia and north-western China. Moves south in winter and regularly observed in Hong Kong.
Status in Japan Rare visitor to inland waters and coastal areas.

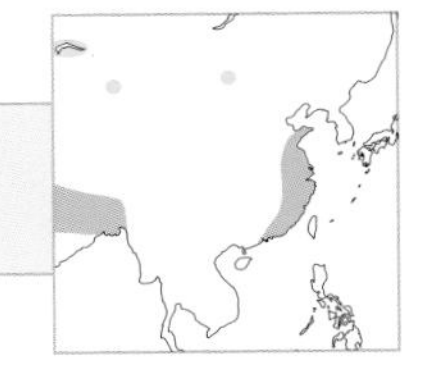

Great White Pelican

Pelecanus onocrotalus 140–175cm WS270cm

Description Very large white waterbird with huge pinkish bill and yellowish bill-pouch. White feathers on body slightly tinged pink, bare pinkish skin around eye and pink legs. Immature has grey-brown back. In flight, white forewings contrast with black primaries and secondaries.
Similar species Dalmatian Pelican has white underwings with grey primaries, greyish legs and dark upper mandible.

Kazuyasu Kisaichi

Adult. November, Chiba

Range Temperate and tropical Eurasia east to Central Asia, also Africa.
Status in Japan Vagrant from central Asia. Occurs on inland lakes and in coastal waters. Presumed aviary escapes are occasionally seen.

Red-tailed Tropicbird

Phaethon rubricauda 96cm WS112cm

Manabu Totsuka

Adult breeding. April, Hawaii, USA

Description Medium-sized seabird with white body, long red tail-streamers, bright red bill and black patch around eye. Juvenile has finely barred upperparts, dark bill and lacks long central-tail feathers.

Similar species White-tailed Tropicbird is smaller and has yellow bill, white tail feathers and conspicuous black upperwing markings. Juveniles can be separated from juvenile White-tailed Tropicbird by their larger size, dark bill and more heavily barred upperparts.

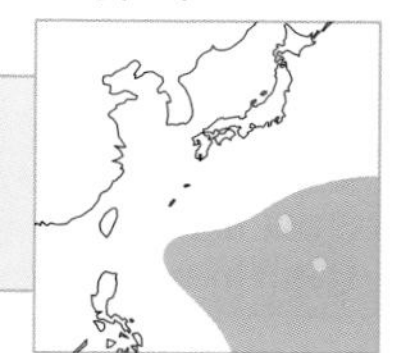

Range Tropical and subtropical Pacific and Indian Oceans.
Status in Japan Rare summer visitor to southern pelagic waters. Breeds on Iwo-jima and Minami-Torishima; rare elsewhere. Most mainland records are birds blown in by typhoons.

White-tailed Tropicbird

Phaethon lepturus 81cm WS92cm

Tokio Sugiyama

Adult breeding. June, Rota Island

Description Medium-sized white seabird with long white tail-streamers, yellow bill and black patch around eye. Conspicuous black markings on upperwings in flight. Juvenile has barred upperparts and pale-yellow bill and lacks long central tail feathers.

Similar species Juveniles can be separated from juvenile Red-tailed Tropicbird by their smaller size, yellow bill and less barred upperparts.

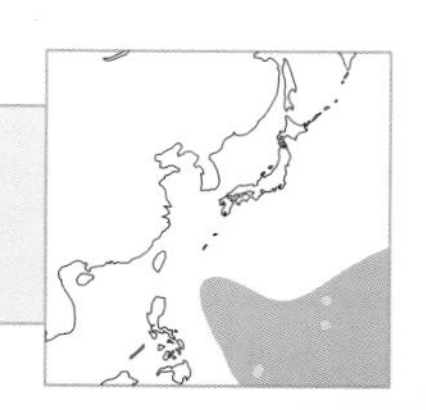

Range Tropical and subtropical Pacific, Indian and Atlantic Oceans.
Status in Japan Rare summer visitor to southern pelagic waters. Recorded from Ryukyu, Iwo, Minami-Torishima and Ogasawara Islands. Few mainland records, mostly after typhoons.

Great Cormorant

Phalacrocorax carbo 82cm WS135cm

Description Large cormorant with glossy-black back, yellow facial skin with round-shaped margin, and white patch at the base of bill. White streaked head and neck and white flank patch in breeding plumage. Immature is browner with variable amounts of white on belly and foreneck.
Similar species Japanese Cormorant has greenish-black back and shorter tail, and triangular-shaped margin at rear of yellow facial skin.

Tadao Shimba
Adult breeding. February, Aichi

Range Widely distributed in temperate and tropical Eurasia, Africa and Australia; also in North America.
Status in Japan Common resident from Honshu south to Kyushu. Moves south in winter. Breeds in colonies in trees. Inhabits coastal areas, estuaries, lakes and rivers.

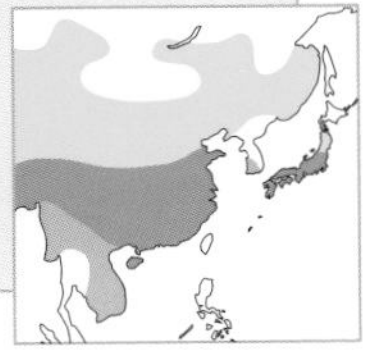

Tadao Shimba
Adult breeding. February, Aichi

Tadao Shimba
Adult non-breeding. February, Aichi

Japanese Cormorant

Phalacrocorax capillatus 84cm WS137cm

Tokio Sugiyama

Adult breeding. February, Shizuoka

Description Large cormorant with greenish-black back, yellow facial skin with triangular-shaped margin, and white patch at base of bill. In breeding plumage has white streaked neck and white flank patch. Immature is browner with variable amount of white on belly and flank.

Similar species Great Cormorant has glossy black back and longer tail. Also has rounded margin to rear of yellow facial skin.

Yoshio Osawa

Adult non-breeding. February, Yamagata

Range Breeds in Japan, Korea, Ussuriland, Sakhalin and southern Kurile Islands. Winters south to eastern China.

Status in Japan Locally common resident on rocky coasts from Kyushu northward. Breeds in colonies on rocky cliffs. This species has been captured and tamed for the ancient fishing technique of catching Japanese trout; now performed only as a tourist attraction in a few places.

Tokio Sugiyama

Immatures. August, Hokkaido

Pelagic Cormorant

Phalacrocorax pelagicus 73cm WS98cm

Description Slender medium-sized cormorant with greenish-black back, small head and slender neck, dark slim bill with red skin at the base. In breeding plumage acquires short crest and white flank patch. Immature is browner overall.
Similar species Both Great and Japanese Cormorants are larger and more robust. Red-faced Cormorant has extensive red facial skin and pale yellow bill.

Kazuyasu Kisaichi

Adult breeding. April, Chiba

Range Widely distributed on both coasts of northern Pacific Ocean.
Status in Japan Locally common resident on rocky coasts from Kyushu northwards. Breeds in colonies on rocky cliffs.

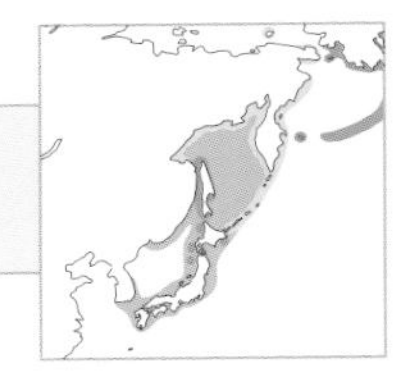

Red-faced Cormorant

Phalacrocorax urile 76cm

Description Similar in size to Pelagic Cormorant but has larger head with paler bill. Distinctive red facial skin, bluish at base of bill. Crest on head in breeding season.
Similar species Pelagic Cormorant has smaller head with darker, slimmer bill.

Kazuyasu Kisaichi

Adult breeding. June, St Paul, Alaska

Range Northern Pacific Ocean south to the Kuril Islands and northern Japan.
Status in Japan Rare winter visitor to rocky coasts on Hokkaido. Breeds in colonies on sea cliffs. Commonly recorded in Hokkaido until 1980s but now only a small population breeds on small islands off eastern Hokkaido.

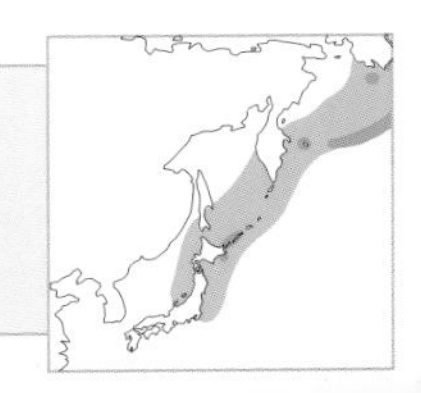

Great Frigatebird

Fregata minor 80–100cm WS220cm

Kazuyasu Kisaichi

Juvenile. November, Chiba

Description Large seabird with very long pointed wings and deeply forked tail. Adult male is all black with distinctive red throat (only visible at close range). Adult female has grey throat, white chest and flanks. Juvenile has orange-brown head with white chest and flanks.

Similar species Lesser Frigatebird is smaller and has conspicuous white patches extending to inner underwings.

Range Widely distributed in tropical Pacific Ocean.

Status in Japan Rare visitor to coastal areas and oceanic islands. Most mainland records are birds blown inshore by typhoons.

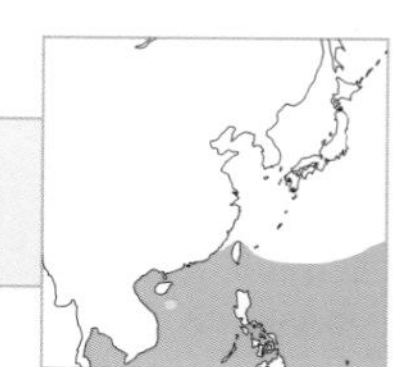

Lesser Frigatebird

Fregata ariel 79cm WS180cm

Norio Yamagata

Juvenile. July, Hokkaido

Description Smaller version of Great Frigatebird. Adult male is entirely black with diagnostic white patch on sides of body extending to inner underwings. Adult female is larger than male with dark head and throat. Both female and juvenile have conspicuous white patches on belly extending to inner underwings.

Similar species See Great Frigatebird.

Range Widely distributed in tropical Pacific Ocean.

Status in Japan Rare visitor to coastal areas and oceanic islands. More frequently recorded than Great Frigatebird. Most mainland records are birds blown inshore by typhoons.

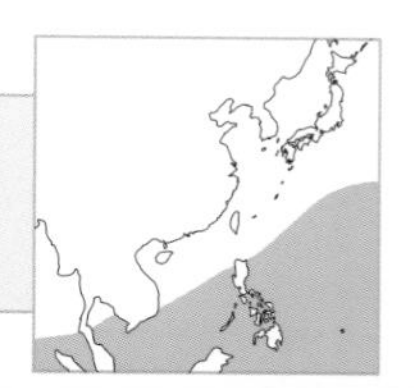

Grey Heron

Ardea cinerea 93cm

Description Large grey heron with yellowish bill, lores and legs. Bill and legs become reddish in breeding season. Immature has browner upper-parts with pale head-stripes.
Similar species Purple Heron is smaller, has rufous head, slimmer neck and long pointed bill.

Tadao Shimba

Adult breeding. April, Aichi

Range Breeds in temperate Eurasia from Europe east to Ussuriland and Sakhalin, south to southern China and south-east Asia; also in Africa. Northern population is migratory.
Status in Japan Common resident from Hokkaido south to Kyushu. Northern population moves south in winter. Breeds in colonies high up in trees. Inhabits coastal mudflats, rice fields and freshwater marshes.

Purple Heron

Ardea purpurea 78.5cm

Description Large heron with rufous head and neck and dark-brown upperparts, distinctive dark stripes from cheeks down sides of neck. Immature is browner, has paler foreneck and lacks dark stripes on neck.
Similar species Grey Heron is larger, has greyish upper-parts and thicker neck. In flight, Purple Heron is more slender and streamlined than Grey Heron.

Tokio Sugiyama

Adult breeding. April, Iriomote-jima

Range Breeds in temperate Eurasia to eastern China and south-east Asia, north to north-eastern China and Ussuriland; also in Africa. Northern population is migratory and rare passage migrant to Korea.
Status in Japan Uncommon resident on Yaeyama Islands, rare elsewhere. Ishigaki-jima is the best place to find this species. Inhabits rice fields and freshwater marshes.

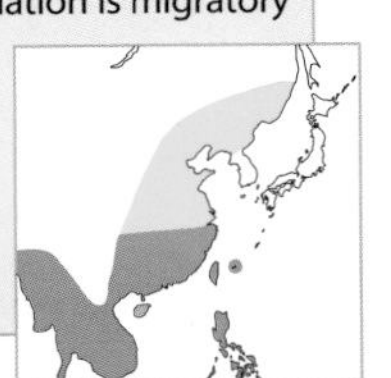

Great Egret

Ardea alba 90cm

Tadao Shimba

Adult breeding *modesta*. May, Aichi

Description Large white egret with very long, slender neck and black legs and feet. In breeding plumage, white plumes on breast and back, black bill and bluish-green lores tinged red. In non-breeding plumage, yellow bill and yellowish lores. Two races in Japan; race *albus* slightly larger with longer legs than *modesta*.
Similar species Intermediate Egret is smaller, has shorter bill and neck; also shorter gape line.

Range Nearly cosmopolitan and widely distributed in temperate and tropical areas. In north-east Asia, breeds from Amur Basin and Ussuriland south to Korea and Japan. Winters in Korea and Japan.
Status in Japan Race *modesta* is an uncommon summer visitor from Honshu south to Kyushu; nominate *alba* is a common winter visitor from Siberia. Lakes, freshwater marshes and tidal flats.

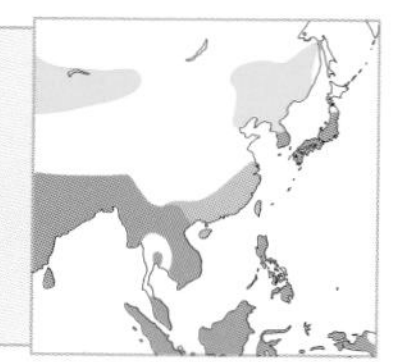

Intermediate Egret

Egretta intermedia 68.5cm

Tadao Shimba

Adult breeding. May, Mie

Other name Yellow-billed Egret
Description Medium-large egret with black legs and feet. In breeding plumage has white plumes on breast and back, black bill and reddish lores. In non-breeding plumage bill becomes yellow with black tip.
Similar species Great Egret is larger with longer bill and neck, also longer gape line extending to rear of the eye. Little Egret is smaller, has shorter neck and yellow feet.

Range Breeds in Asia from Indian subcontinent east to eastern China, south to Australia, also in Africa. Common summer visitor to Korea.
Status in Japan Common summer visitor from Honshu south to Kyushu. Some winter from southern Kyushu southwards. Breeds in colonies with other egrets and inhabits rice fields and freshwater marshes.

Chinese Egret

Egretta eulophotes 65cm

Description Medium-sized white egret with black legs and yellow feet. In breeding plumage has shaggy crest on head, yellow bill and blue lores. In non-breeding plumage the bill becomes black and lores are yellowish-green, as are legs.
Similar species Little Egret has black bill and yellow to reddish lores. Difficult to separate from Little Egret in non-breeding plumage but larger with greenish legs and feet.

Shigeo Ozawa

Adult breeding. May, Tsushima

Range Breeds in northern Korea and eastern China and winters south to Philippines and Malay Peninsula.
Status in Japan Internationally endangered. Rare visitor to coastal mudflats and estuaries. Mostly recorded in spring and summer. Sightings increasing.

Tokio Sugiyama

Adult breeding. April, Tsushima

Tadao Shimba

Adult breeding. May, Hegura-jima

Little Egret

Egretta garzetta 61cm

Tadao Shimba

Adult breeding. April, Aichi

Description Medium-sized white egret with long black bill, black legs, yellow feet and yellow lores. In breeding plumage has two long plumes on nape, upcurved plumes on back and reddish lores. Lores yellow in non-breeding plumage. Juvenile is similar to non-breeding adult, but has yellowish legs.

Similar species Intermediate Egret is larger, has black feet. Adult breeding Chinese Egret has yellow bill, blue lores and shaggy crest, and greenish legs and feet in non-breeding plumage.

Range Breeds in temperate Eurasia east to eastern China, south to Australia, also in Africa. Common resident in Korea.

Status in Japan Common resident from Honshu south to Kyushu, rare in Hokkaido. Breeds in colonies with other egrets. Northern population moves south in winter. Mainly inhabits rivers, ponds and rice fields.

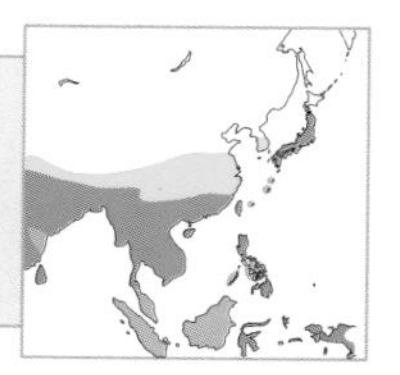

Pacific Reef Egret

Egretta sacra 62.5cm

Tadao Shimba

Adult. December, Okinawa-Hontou

Description Medium-sized egret with two distinct colour morphs; in both morphs bill and leg colour is variable. White morph normally has yellow bill, greenish-yellow lores, greenish-yellow legs and yellow feet. Dark morph is similar with slate-grey plumage.

Similar species White morph can be separated from both Little Egret and Chinese Egret by colour of legs and longer, thicker bill.

Range Resident in coastal areas from northern Australia north through south-east Asia to Taiwan, Korea and Japan.

Status in Japan Uncommon resident on rocky coasts from central Honshu southwards. Dark morph is more common in northern areas; the morphs occur in more equal proportions on Ryukyu Islands.

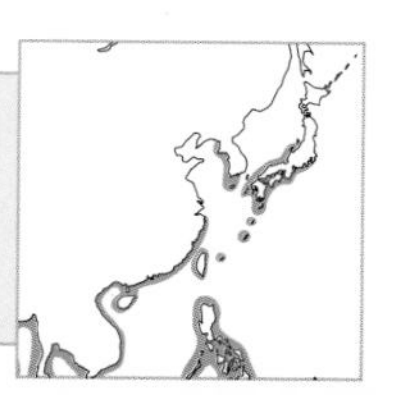

Chinese Pond Heron

Ardeola bacchus 45cm

Description In breeding plumage has rusty head, nape and breast, dark bluish-grey back and dark-tipped yellowish bill. In non-breeding plumage has brown back, heavily streaked dark brown on head and breast. In flight, conspicuous white wings contrast with dark back.
Similar species Separated from all other small herons in the region by all-white wings contrasting with dark back.

Mitsuo Imai

Adult breeding. May, Hegura-jima

Range Breeds in eastern Asia north to Jilin in north-eastern China. Northern population is migratory.
Status in Japan Rare passage and winter visitor to rice fields, ponds and freshwater marshes. Frequently observed on Ryukyu Islands.

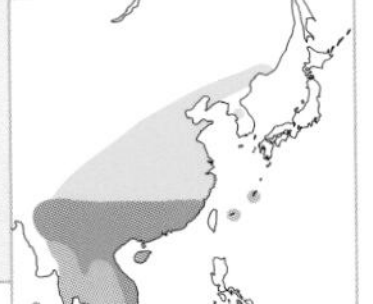

Tadao Shimba

Adult non-breeding. October, Hegura-jima

Cattle Egret

Bubulcus ibis 50.5cm

Description Small, stocky white egret with short neck and short yellow bill; black legs and feet. Unmistakable in breeding plumage with orange head, breast and back; yellow lores become reddish. In non-breeding plumage becomes white overall, but some individuals have yellowish tinge.
Similar species Both Little Egret and Intermediate Egret are larger and have longer bills and necks.

Tadao Shimba

Adult breeding. May, Mie

Range Nearly cosmopolitan and widely distributed in temperate and tropical areas. In north-east Asia, summer visitor to Korea and Japan.
Status in Japan Common summer visitor from Honshu to Kyushu, uncommon on Hokkaido. Winters mainly in southern Japan, common on Ryukyu Islands. Breeds in colonies with other egrets and frequents rice fields and grassy fields. Often follows tractors in rice fields to search for food.

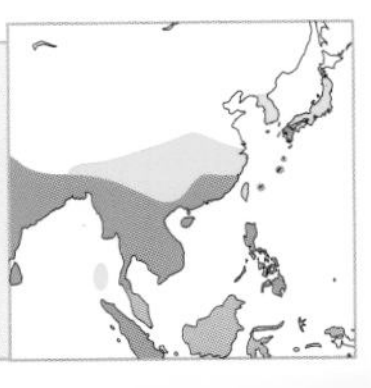

Striated Heron

Butorides striata 52cm

Teruaki Ishii

Adult. June, Kyoto

Description Small heron with black crown, greenish-grey upperparts and pale grey underparts; black bill and yellow legs. Juvenile has dark-brown underparts with white spots on wings.
Voice Loud squeaking *querhuuh*.
Similar species Black-crowned Night Heron is larger with more stocky body and white underparts.

Range Widely distributed in tropical and temperate areas. In north-east Asia, breeds from Amur Basin and Ussuriland south to Korea and Japan. Winters mainly in south of its range.
Status in Japan Uncommon summer visitor from Honshu to Kyushu. Breeds in small colonies high up in trees. Inhabits mainly rivers and streams and feeds mostly on fish.

Black-crowned Night Heron

Nycticorax nycticorax 57.5cm

Tadao Shimba

Adult breeding. April, Aichi

Description Medium-sized stocky heron with black crown and back, greyish-white underparts, black bill and yellow legs. Acquires long white plumes on crown in breeding season. Juvenile brown, heavily streaked below with white spots on back and wing coverts.
Voice Deep loud *gwark* often heard at night.
Similar species Striated Heron is smaller and slimmer, with greenish-grey upperparts and greyish underparts.

Range Widely distributed in temperate and tropical Eurasia; also in Africa and Americas. In north-eastern Asia, breeds in north-eastern China, Korea and Japan. Northern population is migratory.
Status in Japan Common resident around lakes, rivers and ponds from Honshu to Kyushu. Moves south in winter. Mainly nocturnal.

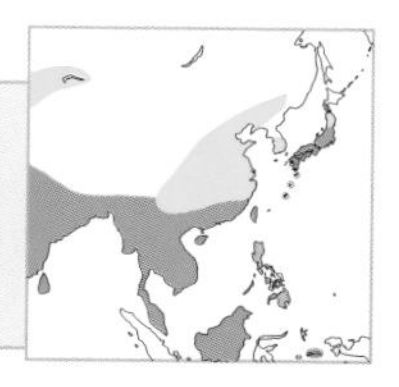

Japanese Night Heron

Gorsachius goisagi 49cm

Description Small heron with reddish-brown upperparts. Dark crown and chestnut head and neck contrast with paler belly; short dark bill and dark green legs. Acquires blue lores in breeding season.
Voice Deep *booh-booh,* calls at night in breeding season.
Similar species Malayan Night Heron is more reddish overall, and has black crown and more extensive blue around lores and eyes, legs are yellowish-green. White tips to upper primaries visible in flight.

Range Breeds in Japan and winters in Philippines and southern China.
Status in Japan Rare summer visitor from Honshu to Kyushu. Breeds in dense forests in low mountains. Secretive and usually detected by booming call. Often observed on Hegura-jima in early May on northbound migration. Numbers declining in recent years and endangered.

Kazuyasu Kisaichi

Adults on nest. June, Chiba

Tokio Sugiyama

Adult. April, Aichi

Tokio Sugiyama

Adult. May, Hegura-jima

Malayan Night Heron

Gorsachius melanolophus 47cm

Tadao Shimba

Adult. April, Ishigaki-jima

Description Small heron with reddish-brown upperparts, black crown and chestnut face and neck. Lores blue or green. In flight, shows white tips to upper primaries. Juvenile is pale greyish-brown with distinctive white spots.
Voice Similar to Japanese Night Heron.
Similar species See Japanese Night Heron.

Tadao Shimba

Adult. April, Ishigaki-jima

Range Southern and eastern Asia. Breeds from India east to South-east Asia, north to Taiwan and southern Ryukyu Islands.
Status in Japan Uncommon resident on Yaeyama Islands; inhabits dense evergreen forests. Often seen feeding on open grassy fields, and normally approachable.

Kenji Takehara

Juvenile. March, Kohama-jima, Yaeyama Islands

Yellow Bittern

Ixobrychus sinensis 36.5cm

Description Smallest bittern in the region. Male has black head, reddish-brown back and pale-brown underparts, yellow bill and greenish legs. Female has reddish-brown head and brown streaks from foreneck to breast. Immature has heavy dark streaks on back and below. In flight, yellowish-brown wing coverts contrast with dark flight feathers.
Voice Muffled and booming *wooh-wooh*, calls at night in breeding season.
Similar species Both female Schrenck's Bittern and female Cinnamon Bittern have white spotted backs.

Akira Yamamoto
Adult ♀ breeding. June, Aichi

Mitsuo Imai
Adult ♂ breeding. June, Chiba

Range Southern and eastern Asia. Breeds from India east to eastern China, north to north-eastern China, Korea and Japan; south to Indonesia. Northern populations are migratory.
Status in Japan Uncommon summer visitor to freshwater marshes, lakes and reed beds. Uncommon winter visitor to Ryukyu Islands.

Teruaki Ishii
Adult ♂ non-breeding. October, Osaka

Schrenck's Bittern

Ixobrychus eurhythmus 40cm

Kazuyasu Kisaichi

Adult ♂. July, Ibaragi

Description Very small. Male has black crown, chocolate-brown face, nape and back contrasting with pale yellowish-brown underparts; single dark streak in the centre of foreneck and breast. Female has dark crown, chestnut-brown back with white spots, heavy dark streaks on underparts. In flight, grey forewings contrast with dark flight feathers.
Similar species Female Cinnamon Bittern has reddish-brown face and back.

Akira Yamamoto

Adult ♂. May, Ibaragi

Range Eastern Asia. Breeds from Amur basin, Ussuriland, Sakhalin south to eastern China. Winters south to south-eastern Asia. Uncommon summer visitor to Korea.
Status in Japan Rare summer visitor to large reed beds from central Honshu northwards. Rare passage migrant elsewhere. A few winter records. Declined significantly since 1980s.

Teruaki Ishii

Adult ♀. November, Osaka

Cinnamon Bittern

Ixobrychus cinnamomeus 40cm

Description Very small. Male has uniform rufous upperparts and paler underparts, yellowish bill and greenish-yellow legs. Female is darker than male, has white spots on back and is heavily streaked from foreneck to breast. Immature has mottled dark-brown upperparts and heavily streaked underparts.

Similar species Yellow Bittern has yellowish-brown back with light brown forewings contrasting with dark brown flight feathers. Female Schrenck's Bittern is darker, with heavy white spots on back.

Range Southern and eastern Asia. Breeds from India east to eastern China, Taiwan and Ryukyu Islands, south to Indonesia. Northern population is migratory.

Status in Japan Common resident from Amami Islands southward. Inhabits rice fields, reed beds and freshwater marshes.

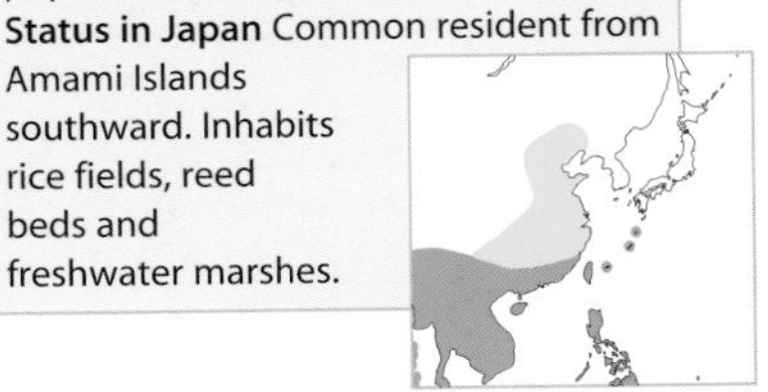

Kazuyasu Kisaichi

Adult ♂. June, Ishigaki-jima

Akira Yamamoto

Adult ♀. August, Miyako-jima

Kazuyasu Kisaichi

Juvenile. October, Ishigaki-jima

Eurasian Bittern

Botaurus stellaris 70cm

Shigeo Ozawa
June, Chiba

Kazuyasu Kisaichi
May, Chiba

Kazuyasu Kisaichi
June, Chiba

Description Large bittern with black crown and yellowish-brown back mottled with black; thick neck with black streaks from foreneck to breast; greenish-yellow bill and legs. Freezes and relies on camouflage when approached.
Voice Muffled, booming *booh-booh* in breeding season.
Similar species Immature Black-crowned Night Heron is smaller with darker brown back and shorter neck.

Range Breeds in temperate Eurasia from Europe east through Siberia to Amur Basin, Ussuriland, Sakhalin and north-east China. Winters south to northern Africa and southern Asia.
Status in Japan Rare resident. Breeds in large freshwater marshes with reed beds in Hokkaido and few places in central Honshu. Winters from Honshu southwards.

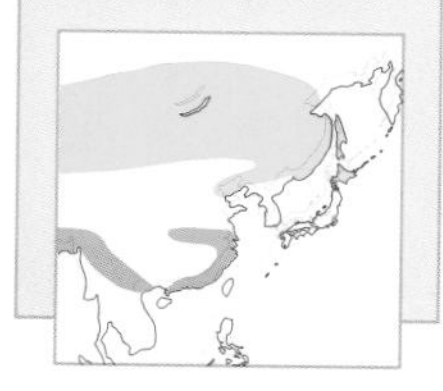

Black Stork

Ciconia nigra 99cm

Description Large black and white waterbird with long red bill and red legs. Glossy-black head, neck and breast contrast with white underparts. Juvenile has brownish head, neck, breast and back with buff-coloured spots on breast; and greyish-green bill and legs.

Shigeo Ozawa

Adult. January, Miyazaki

Range Breeds in temperate Eurasia from central Europe east to northern China and far-eastern Russia. Winters south to northern Africa, India and southern China.

Status in Japan Rare passage migrant and winter visitor to cultivated fields, freshwater marshes and rivers.

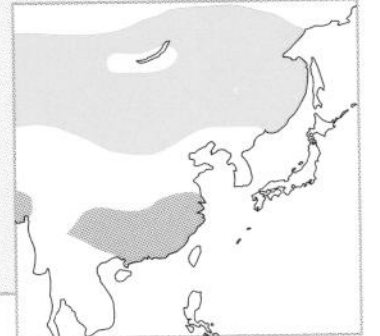

Oriental Stork

Ciconia boyciana 112cm

Description Large white waterbird with long black bill, pale yellow eye and long red legs. In flight, black flight feathers contrast with white forewings.

Similar species Siberian Crane has red face and white secondaries.

Tadao Shimba

Adult. December, Shizuoka

Range Breeds in Amur Basin, Ussuriland and north-east China and winters in eastern China. Rare winter visitor to Korea.

Status in Japan Common resident until 19th century. Seriously declined during 20th century due to habitat destruction and use of agricultural pesticides. Wild population exterminated by 1970s but recovery programme in place since 1960s. Several birds released to the wild in 2005 in Hyogo. Some wild birds occur occasionally, mainly in winter, on rice fields, ponds and rivers.

Black-headed Ibis

Threskiornis melanocephalus 68cm

Akira Yamamoto

Immature. January, Kyoto

Description Large white waterbird with distinctive long, black, downcurved bill and black legs. Adult has bare black skin on head and upper neck. Immature has dark-grey feathered head with dark tips to primaries.
Similar species Both Eurasian and Black-faced Spoonbills have spatulate bills and feathered heads.

Range Breeds mainly in Indian subcontinent and north-eastern China. Winters south to south-eastern Asia.
Status in Japan Rare visitor to cultivated fields, rivers and freshwater marshes. More records in western Japan.

Eurasian Spoonbill

Platalea leucorodia 86cm

Tokio Sugiyama

Juvenile. November, Shizuoka

Description Large white waterbird with distinctive large spatulate bill and long black legs. Acquires yellow crest and yellow tip to bill in breeding plumage. Immature has pink tinged dark bill and black tips to flight feathers.
Similar species Black-faced Spoonbill has black bill and extensive bare black skin on forehead and around eye.

Range Breeds in temperate Eurasia from southern Europe east to Ussuriland and north-eastern China; also in India and Africa. Winters south to tropical Africa and southern Asia. Rare winter visitor to Korea.
Status in Japan Rare winter visitor to freshwater marshes, coastal mudflats and freshwater marshes. Small number regularly winter in Kyushu.

Black-faced Spoonbill

Platalea minor 73.5cm

Description Similar to Eurasian Spoonbill, but bill entirely black and large area of exposed black skin on forehead and around eye. Acquires yellow crest and breast in breeding plumage. Immature has pinkish bill, black outer primaries and dark trailing edges to primaries and secondaries.
Similar species Eurasian Spoonbill has less exposed bare skin in front of eye.

Akira Yamamoto

Adult breeding. March, Aichi

Akira Yamamoto

Juvenile. November, Aichi

Range Restricted to far-eastern Asia and internationally endangered. Breeds mainly on small islands off west coast of Korea and winters in south-eastern China, Taiwan and southern Japan.
Status in Japan Rare winter visitor to freshwater marshes, coastal mudflats and freshwater marshes. Small flocks regularly winter in northern Kyushu and Okinawa-Hontou.

Akira Yamamoto

Adult breeding. March, Aichi

Mute Swan

Cygnus olor 152cm

Tadao Shimba

Adult. May, Mie

Description Large swan with reddish-orange bill and distinctive black knob at the base of bill. Juvenile is greyish-brown overall; has grey bill with black at base.

Range Common resident in Europe, and breeds sparsely from Caspian Sea area east to Transbaikalia and Mongolia. Partially migratory and rare winter visitor to Korea.
Status in Japan Introduced to parks and gardens with escaped birds breeding freely; some regularly migrate from Hokkaido to Ibaragi. Only one official record of a wild bird, in 1933 on Hachijo-jima.

Trumpeter Swan

Cygnus buccinator 152cm

Mitsuo Imai

Adult. December, Miyagi

Description Large swan with black bill and black lores. Juvenile is greyish-brown overall with dark bill and pale tip.
Similar species Tundra Swan of race *columbianus* also has black bill, but smaller, and has small yellow spot at the base of bill.

Range Breeds mainly in central Alaska and winters along Pacific coasts of North America.
Status in Japan Vagrant. Single bird, presumed to be same individual, spent two consecutive winters in Miyagi and Iwate (1991–92, 1992–93); also in 2006 in Hokkaido.

Whooper Swan

Cygnus cygnus 140cm

Description Large swan with dark-tipped yellow bill. Juvenile is dark greyish-brown overall and has dark-tipped pinkish bill.
Similar species Tundra Swan is smaller, shorter-necked and has slightly shorter bill with smaller yellow patch.

Range Breeds across northern Eurasia from Scandinavia east to Kamchatka Peninsula, south to Sakhalin, Amur Basin and north-eastern China. Moves south in winter and common winter visitor to central China, Korea and Japan.
Status in Japan Locally common winter visitor to lakes, rivers and coastal waters from central Honshu northward. Often tame and can usually be seen at close range.

Tadao Shimba

Adult. December, Hokkaido

Tadao Shimba

Juvenile. December, Hokkaido

Tadao Shimba

Adult. December, Hokkaido

Tundra Swan

Cygnus columbianus 120cm

Tadao Shimba

Adult *bewickii*. February, Nagano

Other name Bewick's Swan
Description Race *bewickii*, medium-sized swan with black bill with yellow patch at the base. Juvenile is dark greyish-brown overall and black bill with pale yellow base. Nominate *columbianus* has mainly black bill with much smaller yellow patch at base.
Similar species See Trumpeter and Whooper Swans.

Tadao Shimba

Juvenile *bewickii*. February, Nagano

Range Breeds on arctic tundra in Eurasia and North America and migrates in autumn to wintering areas. Common winter visitor to eastern China, Korea and Japan.
Status in Japan Locally common winter visitor to lakes, rivers and coastal waters from Honshu northwards. Race *columbianus* from North America is rare, but regularly observed.

Tadao Shimba

Adult *columbianus* with *bewickii* in background. February, Nagano

Swan Goose

Anser cygnoides 87cm

Description Large goose with long neck and long black bill with white line at the base. Greyish-brown back, brown belly and white undertail-coverts. Brown crown and hind-neck contrast with pale chin, cheeks and foreneck. Orange legs.

Tokio Sugiyama

Adult. December, Fukuoka

Range Breeds from the Altai east to Amur Basin, Ussuriland and north-eastern China. Winters mainly in eastern China and uncommon winter visitor to Korea.

Status in Japan Rare winter visitor to rice fields and freshwater marshes. Most often observed in western Japan. Flocks of hundreds used to visit Tokyo Bay until 1950s.

Tokio Sugiyama

Adult. December, Fukuoka

Tokio Sugiyama

Adult, with Bean and Greater White-fronted Geese. January, Yamaguchi

Bean Goose

Anser fabalis 85cm

Tadao Shimba

Adult *middendorffi*. February, Nagano

Description Large goose with dark-brown head and neck. Greyish-brown back and under-parts, dark bill with orange band and white uppertail-coverts, orange legs. Two races visit Japan in winter; *middendorffii* has a long pointed bill; *serrirostris* has a short bill. **Similar species** Greater White-fronted Goose has pink bill with white patch at the base.

Tadao Shimba

Juvenile *middendorffi*. November, Shiga

Yoshio Osawa

Adult *serrirostrisis*. February, Shimane

Range Breeds across northern Eurasia and moves south in winter. Race *middendorffii* breeds in the taiga zone of eastern Siberia, south to Lake Baikal and northern Amur Basin; *serrirostris* breeds on arctic tundra in eastern Siberia. Both races winter in China, Korea and Japan. **Status in Japan** Locally common winter visitor to lakes, freshwater marshes and rice fields from Honshu northward. Race *middendorffii* visits coastal areas of Sea of Japan and inhabits shallow lakes and freshwater marshes; *serrirostrisis* visits Pacific coast of northern Honshu and feeds mainly in rice fields.

Greater White-fronted Goose

Anser albifrons 72cm

Description Greyish-brown goose with pink bill and orange legs, white patch at the base of bill and white undertail-coverts. Adult has irregular black patches on belly. Juvenile slightly browner and lacks white patch at the base of bill.
Similar species Lesser White-fronted Goose is smaller, and has shorter bill and yellow eye-ring; also white patch at the base of bill extends on to forehead.

Tadao Shimba

Adult. November, Miyagi

Range Breeds on arctic tundra in Eurasia, North America and western Greenland. Moves south in winter. Common winter visitor to Korea.
Status in Japan Common and locally abundant winter visitor to northern Japan; rare elsewhere. Rice fields and freshwater marshes.

Tadao Shimba

Juvenile. November, Miyagi

Manabu Totsuka

Adult. November, Miyagi

Lesser White-fronted Goose

Anser erythropus 58.5cm

Akira Yamamoto

Adult. November, Miyagi

Description Greyish-brown goose with short pink bill and orange legs. White patch at the base of bill extends to crown; yellow eye-ring and white undertail-coverts. Adult has irregular black patches on belly. Juvenile is dark brown overall and lacks white patch at the base of bill.

Similar species Greater White-fronted Goose is larger, has longer bill, with less extensive white blaze at base and lacks yellow eye-ring.

Tokio Sugiyama

Juvenile. December, Shimane

Range Breeds on arctic tundra in Eurasia from northern Scandinavia east to north-eastern Russia. Winters from south-eastern Europe east to southern China.

Status in Japan Rare winter visitor to rice fields and freshwater marshes. Small flocks regularly visit Miyagi and Shimane and often found among flocks of Greater White-fronted Goose.

Snow Goose

Anser caerulescens 67cm

Kazuyasu Kisaichi

Adult white morph. February, Aomori

Description White morph is snow-white with pink bill and legs. In flight, distinctive black primaries contrast with white forewings and secondaries. Blue morph is dark grey with white head and neck. Juvenile is greyish-brown and has dark bill.

Range Breeds on Wrangel Island off northern coast of Siberia and in arctic region of North America. Winters south to Gulf Coast of North America.

Status in Japan Rare winter visitor to lakes and rice fields in northern Japan. Common winter visitor until 19th century.

Emperor Goose

Chen canagica 66cm

Description Small, stocky goose with short, thick neck. White head and hind-neck, small pinkish bill and orange legs; face often stained orange in breeding plumage. Black breast, bluish-grey back and belly with black-and-white edges. Juveniles have dark head and bill.
Similar species Blue–morph Snow Goose is larger, has white head and upper neck, pinkish legs and light grey wing-coverts.

Mike Boylan

Adults. Adak Island, Alaska

Range Breeds on tundra on Chukotka and in north-west Alaska. Winters in coastal areas south to Aleutians and southern Alaska.
Status in Japan Vagrant. One wintered on rice fields among Greater White-fronted Geese in Miyagi in mid-1960s.

Donna Dewhurst

Adults and young. Yukon Delta, Canada

Tuomo Jaakkonen

Adult. June, Chukotka, Russia

Tuomo Jaakkonen

Adult. June, Chukotka, Russia

Greylag Goose

Anser anser 84cm

Kenji Takehara

Adult. December, Okinawa-Hontou

Description Large greyish-brown goose with pink bill and legs, greyish belly and white undertail-coverts. Juvenile is duller. In flight, grey forewings contrast with dark hindwings and back.

Similar species Greater White-fronted Goose has darker belly and white patch at the base of bill. Bean Goose is darker and has dark bill with orange band.

Tokio Sugiyama

Juveniles. November, Aichi

Range Widely distributed in Eurasia and breeds from Europe east to the Amur Basin, northern Mongolia and northern China. Moves south in winter.

Status in Japan Rare passage and winter visitor to rice fields, rivers and shallow waters.

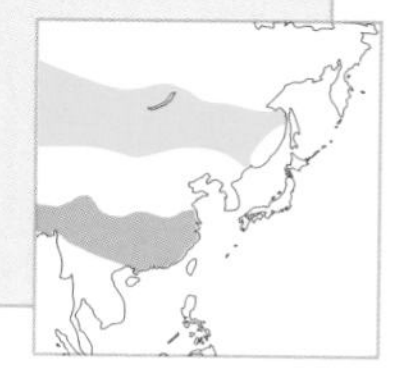

Kazuyasu Kisaichi

Adult. March, Miyagi

Cackling Goose

Branta hutchinsii 67cm

Description Diagnostic white cheeks and neck-ring contrast with black head and neck, barred light-brown belly and white undertail-coverts, dark greyish-brown back with buff-coloured edges and dark primaries.
Similar species Brent Goose lacks white chin-strap.
Note Formerly considered a subspecies of Canada Goose *B. canadensis*.

Range Breeds in northern North America and winters south to northern Mexico and Gulf Coast.
Status in Japan Rare winter visitor to northern Japan. Small numbers regularly visit Miyagi in winter. Two races recorded; mainly *leucopareia* from Aleutian Islands; race *minima* also observed. Common winter visitor until early 20th century and flocks of hundreds used to winter in northern Japan.

Kazuyasu Kisaichi

Adult *leucopareia*. November, Ibaragi

Kazuyasu Kisaichi

Adult *leucopareia*. November, Ibaragi

Kazuyasu Kisaichi

Adult *minima*. February, Chiba

Brent Goose

Branta bernicla 61cm

Tadao Shimba

Adult. December, Hokkaido

Description Race in our region is *nigricans*, the Black Brant, which is sometimes considered a separate species. Small goose with black head, neck, breast and back, white patch at the sides of neck, black legs, white belly with black bars and white undertail-coverts. Juvenile has browner back with white edges to back feathers.
Similar species Cackling Goose has white cheeks and neck–ring.

Tokio Sugiyama

Juvenile. January, Shizuoka

Range Breeds on arctic tundra of North America and Eurasia and winters in temperate regions to the south.
Status in Japan Locally common winter visitor to rocky shores and shallow bays in northern Japan. Rarely occurs on inland waters.

Tokio Sugiyama

Juvenile. January, Shizuoka

Ruddy Shelduck

Tadorna ferruginea 63.5cm

Description Large, orange-brown duck with dark bill and legs. In breeding plumage male acquires black neck-ring, female has paler face. In flight, white forewings contrast with black primaries and secondaries.

Range Breeds sparsely in temperate Eurasia east to Transbaikalia, Mongolia, Amur Basin and north-east China. Moves south in winter and is an uncommon winter visitor to Korea.

Status in Japan Rare winter visitor to rice fields, rivers and shallow waters. Most frequently observed in western Japan.

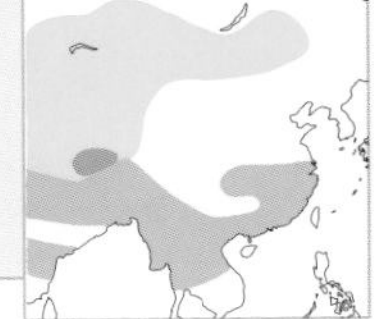

Kazuyasu Kisaichi

Adult ♀. March, Tokyo

Common Shelduck

Tadorna tadorna 62.5cm

Description Unmistakable large duck. Greenish-black head, white back with dark shoulders, white underparts with broad chestnut breast-band extending to back, red bill and legs. Male has red knob at the base of bill in breeding season. In flight, white forewings contrast with dark primaries and secondaries.

Range Breeds across Eurasia from Europe east to Mongolia, Transbaikalia, Amur Basin and north-eastern China. Moves south in winter and common winter visitor to Korea.

Status in Japan Locally common winter visitor to coastal mudflats in northern Kyushu; rare elsewhere.

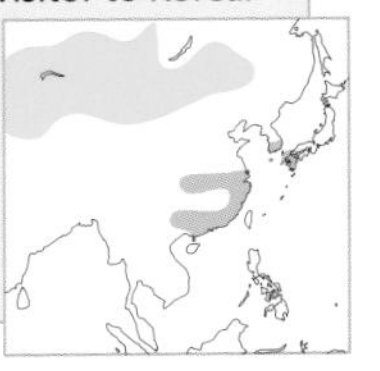

Akira Yamamoto

Adult ♂ breeding. May, Aichi

Mandarin Duck

Aix galericulata 45cm

Tadao Shimba

Adult ♂ breeding. February, Aichi

Description Breeding male unmistakable. Broad white supercilium, long chestnut hackles and red bill, purplish breast with two white lines, greenish-purple crown and crest, bright orange flanks and distinctive fan-shaped tertials. Female is greyish-brown with white eye-ring and white stripe behind eye. Eclipse male as female but bill pinkish.

Manabu Totsuka

Adult ♀. February, Aichi

Range Breeds in Amur Basin, Ussuriland and Sakhalin, south to Korea and Japan. Northern population is migratory and winters south to southern China and Taiwan. Uncommon resident in Korea.

Status in Japan Common resident from Kyushu northward. Breeds in holes in trees in mountain forests near streams and moves to lower altitudes in winter. Often seen in city parks and gardens in winter. Summer visitor to Hokkaido.

Tokio Sugiyama

Adults. February, Aichi

Eurasian Wigeon

Anas penelope 48cm

Description Breeding male has reddish-brown head with distinctive golden-brown forehead and bluish-grey bill with dark tip, pinkish-brown breast, grey back and flanks, black undertail-coverts with white band at the side. In flight, white forewings and green speculum on upperwings are visible. Female is brown overall, with bluish bill with dark tip. In flight, green speculum with white leading edges and grey axillaries visible. Eclipse male has head, neck and body more rufous. **Similar species** Male American Wigeon has distinctive broad green eye-patch and white forehead. Female American Wigeon is greyer with white axillaries.

Tadao Shimba

Adult ♂ breeding. February, Aichi

Tadao Shimba

Adult ♀. February, Aichi

Range Breeds across northern Eurasia from Scandinavia east to Kamchatka Peninsula, south to north-east China and Sakhalin. Winters south to southern China and Taiwan.
Status in Japan Common winter visitor to lakes, rivers and shallow coastal waters.

American Wigeon

Anas americana 48cm

Description Breeding male has white forehead and crown, and broad green eye-patch extending to nape. Buff-coloured cheeks and neck, grey back, pinkish-brown breast and flanks, black undertail-coverts with white band in front. In flight, white forewings and green speculum on upperwings visible. Female similar to Eurasian Wigeon but has greyish head and white axillaries. In flight, green speculum with white edges visible. Eclipse male resembles female but forewings white.

Akira Yamamoto

Adult ♂ breeding. January, Aichi

Range Breeds in northern North America and winters mainly along coasts. Rare winter visitor to Korea.
Status in Japan Rare winter visitor to lakes, rivers and shallow coastal waters. Normally found amongst flocks of Eurasian Wigeon. Most males have yellow foreheads indicating hybridisation with Eurasian Wigeon.

Mallard

Anas platyrhynchos 64cm

Tadao Shimba

Adult ♂ breeding. February, Aichi

Description Breeding male unmistakable. Greenish-black head and chestnut breast separated by white collar; greyish-brown back and silvery-grey flanks, yellow bill and orange feet. Female is mottled brown, with pale face, orange bill with dark spot and orange feet. Purplish-blue speculum on upperwings in flight. Eclipse male resembles female but bill yellow. **Similar species** Spot-billed Duck is similar to female, but has dark eye-stripe and yellow-spotted dark bill.

Tadao Shimba

Adult ♀. November, Aichi

Range Abundant and widespread in northern hemisphere. In north-eastern Asia breeds mainly from north-eastern China northward and winters in Korea and Japan. **Status in Japan** Abundant winter visitor to lakes, rivers and shallow waters. A few breed in Hokkaido and Honshu.

Tadao Shimba

Adult ♂ breeding. February, Aichi

Falcated Duck

Anas falcata 48cm

Description Breeding male has distinctive glossy-green head with finely speckled grey body. Pale yellow neck with black collar, long tertials droop over tail, black undertail-coverts with wide pale yellow patch. Female is mottled dark-brown with dark bill. In flight, green speculum edged by white visible. Eclipse male is darker than female, with grey forewing.

Similar species Female Gadwall has yellow bill and feet and shows small white speculum in flight. Female Eurasian Wigeon is reddish-brown overall with bluish-grey bill.

Tadao Shimba

Adult ♂ breeding. February, Tokyo

Tadao Shimba

Adult ♀. April, Shizuoka

Range Breeds from the Yenisey east to Okhotsk coast and Kamchatka Peninsula, south to northern China, Ussuriland and Sakhalin. Winters in eastern China, Korea and Japan.

Status in Japan Uncommon winter visitor to lakes, rivers and shallow coastal waters. Some breed in Hokkaido.

Gadwall

Anas strepera 50cm

Description Breeding male has greyish-brown head, dark bill, finely speckled greyish breast and flanks; greyish-brown back, black undertail-coverts and yellowish-orange legs. Small white speculum. Female is mottled dark-brown overall with yellow bill and yellowish-orange feet; small white speculum. Eclipse male resembles female but tertials grey.

Similar species Female Mallard is larger with white tail and large blue speculum. Female Falcated Duck has darker bill.

Tadao Shimba

Adult ♂ breeding. February, Nagano

Range Breeds in temperate Eurasia and North America. In north-east Asia, breeds sparsely in Kamchatka Peninsula, Sakhalin and north-eastern China. Moves south in winter and common winter visitor to Korea.

Status in Japan Uncommon winter visitor to lakes, rivers and shallow waters. Some breed in northern Japan.

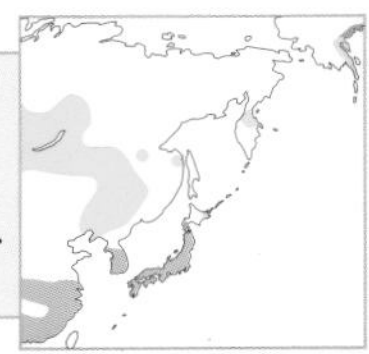

Baikal Teal

Anas formosa 40cm

Tadao Shimba

Adult ♂ breeding. February, Nagano

Description Breeding male unmistakable yellow, green and black facial pattern. Reddish-brown breast, brown back, long drooping black–and–white striped scapulars, bluish-grey belly and flanks. Female is mottled brown overall, dark bill with white spot at the base. In flight shows green speculum on upperwings. Eclipse male as female but more rufous. **Similar species** See Common Teal; also Garganey.

Tadao Shimba

Adult ♀. February, Nagano

Range Breeds in eastern Siberia east to Kamchatka Peninsula, south to Lake Baikal and Amur Basin. Winters mainly in Korea; also in eastern China and Japan. **Status in Japan** Uncommon winter visitor to lakes, rivers and shallow coastal waters. Most frequently observed in western Japan.

Akira Yamamoto

Adult ♂ breeding. January, Tokyo

Common Teal

Anas crecca 37.5cm

Description Breeding male has dark chestnut head with dark glossy-green patch; finely speckled buff-coloured breast, greyish-brown back and grey flanks with distinct white line on scapulars; black-bordered yellow patch at stern. Female has mottled dark-brown body with dark eye-stripe. In flight, male has white broad line on scapulars and green speculum on upperwings; female has green speculum edged by white. Eclipse male resembles female.
Similar species Female Garganey has dark facial stripes, pale supercilium and white spot at the base of bill. Female Baikal Teal is slightly larger; also has pale spot at the base of bill.

Tadao Shimba

Adult ♂ breeding *crecca*. March, Aichi

Range Breeds across northern Eurasia from Europe east to Okhotsk coast, Kamchatka and Sakhalin, south to northern China and northern Japan. Common winter visitor to Korea.
Status in Japan Common winter visitor to ponds, marshes and rivers. Prefers to stay near cover. Few breed from central Honshu northward. The Nearctic form *carolinensis* is often treated as a separate species, the Green-winged Teal; a rare winter visitor, the male has a distinct vertical stripe in front of the wing; female inseparable from Common Teal.

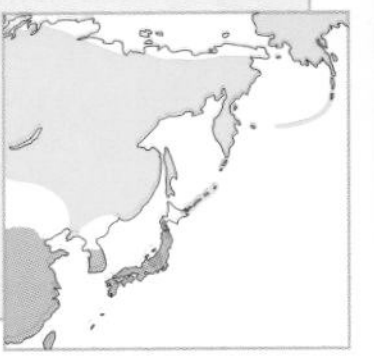

Tadao Shimba

Adult ♀ *crecca*. March, Aichi

Norio Yamagata

Adult ♂ breeding *carolinensis*. January, Aichi

Spot-billed Duck

Anas zonorhyncha 60.5cm

Mitsuo Imai

Adult ♀ (l) and ♂ (r). March, Fukishima

Description Large, mottled, dark-brown duck with pale face and dark eye-stripe. Dark-brown back contrasts with white tertials, dark bill with yellow spot at the tip. Sexes similar, but male slightly darker. In flight, purplish-blue speculum on upperwings and white underwing-coverts.

Similar species Female Mallard has orange bill and lacks dark stripe on face.

Tadao Shimba

Adult ♀. May, Aichi

Range Breeds in temperate and tropical Asia, north to Transbaikalia, Amur Basin and Sakhalin. Northern population moves to warmer areas in winter. Common resident in Korea.

Status in Japan Abundant resident on freshwater lakes and rivers throughout Japan; often breeds in city parks. Northern population moves south in winter.

Tokio Sugiyama

Adult ♀ and ducklings. July, Aichi

Northern Pintail

Anas acuta ♂ 75cm ♀ 53cm

Description A slim, elegant duck. Breeding male has grey body with very long pointed tail. Chocolate-brown head and hindneck, conspicuous white line running from the side of the neck to breast, bluish-grey bill and legs, black undertail-coverts bordered by pale yellow band. Female is mottled greyish-brown, with bluish-grey bill and legs. In flight, rusty-brown speculum with buff-coloured leading-edges and white trailing-edges visible. Eclipse male is similar to female but scapulars greyer.

Similar species Both female Mallard and Gadwall are stockier, have yellowish bills and different upperwing patterns.

Range Breeds across Eurasia from Europe east to Kamchatka Peninsula, south to Amur Basin and Sakhalin; also in North America. Moves south in winter and common winter visitor to Korea.

Status in Japan Common winter visitor to lakes, rivers and shallow coastal waters.

Tadao Shimba

Adults. February, Nagano

Tadao Shimba

Adult ♀. February, Nagano

Tadao Shimba

Adult ♂ breeding. December, Tokyo

Garganey

Anas querquedula 38cm

Akira Yamamoto ©

Adult ♂ breeding. April, Aichi

Akira Yamamoto ©

Adult ♀. April, Aichi

Akira Yamamoto ©

Juveniles. August, Aichi

Description Breeding male has distinctive broad white supercilium, curving downwards on hindneck. Dark-brown face, breast and back with long drooping black and white scapulars, white belly. Female is similar to female Common Teal but has pale supercilium, dark stripes on face and pale spot at the base of bill. In flight, dark grey forewings contrast with dark primaries and white-edged greenish speculum.

Similar species Both female Common and Baikal Teal lack dark facial stripes and pale supercilia.

Range Breeds in Eurasia from Europe east through central Asia to Kamchatka Peninsula, Sakhalin and northern Japan. Winters south to northern Africa, Indian subcontinent and South-east Asia.

Status in Japan Uncommon passage migrant to fresh-water ponds and marshes. A few breed in northern Japan and some winter on Ryukyu Islands.

Blue-winged Teal

Anas discors 36–41cm

Description Breeding male has bluish-grey head with distinctive, crescent-shaped, white patch in front of eye. Female is mottled brown, with white eye-ring and pale patch at the base of bill. In flight, both sexes show pale-blue upperwing-coverts. Eclipse male is as female but shows broader white covert-band in flight.
Similar species Female Garganey has white supercilium and greyish-brown wing-coverts. Female Baikal and Common Teals have dark upperwing-coverts and smaller bills.

Norio Yamagata

Adult ♂ breeding. February, Gifu

Tadao Shimba

Adult ♀. May, Ontario, Canada

Range Breeds in northern and central North America and winters from southern North America southward.
Status in Japan Vagrant. One record: a male in 1996 in Gifu.

Northern Shoveler

Anas clypeata 50cm

Description Medium-sized duck with unmistakable spatulate bill. Breeding male has greenish-black head, dark back, white breast and chestnut belly. In flight, pale-blue upperwing-coverts and green speculum edged by white visible. Female has mottled brown body and orange-brown bill. Blue-grey upperwing-coverts visible in flight. Eclipse male is as female but head darker, and retains pale blue upperwing-coverts.

Akira Yamamoto

Adult ♂ breeding. January, Aichi

Range Breeds in temperate Eurasia and North America, moves south in winter. In north-east Asia, common summer visitor in north; common winter visitor in Korea and Japan.
Status in Japan Common winter visitor to ponds, freshwater marshes and shallow coastal waters.

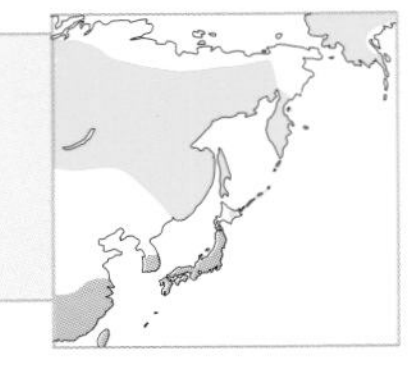

Red-crested Pochard

Netta rufina 56cm

Teruaki Ishii

Adult ♂ breeding. December, Kanagawa

Teruaki Ishii

Adult ♀. January, Osaka

Teruaki Ishii

Adult ♂ breeding. December, Kanagawa

Description Large diving duck. Breeding male has distinctive red bill, reddish-orange head, black neck and breast; greyish-brown back with white shoulder patch, white flanks with irregular brown bars. Female has dark-brown cap, pale cheeks, dark bill with pink tip, greyish-brown back, pale-brown breast and belly. In flight, shows prominent white wing-bars. Eclipse male resembles female but bill is red.

Range Breeds locally in temperate Eurasia from southern Europe east to Mongolia and Transbaikalia. Winters south to northern Africa and southern Asia.

Status in Japan Rare winter visitor to lakes, ponds and rivers. Increasing sightings in recent years and small number regularly visit Biwa Lake in Shiga.

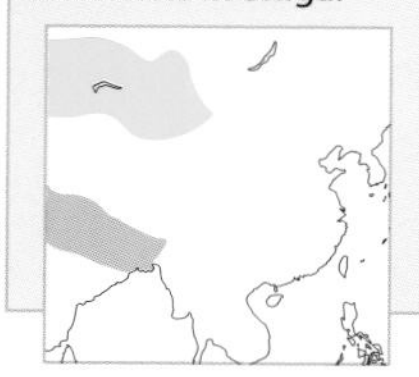

Common Pochard

Aythya ferina 45cm

Description Medium-sized diving duck. Breeding male has reddish-brown head, red eyes and dark-blue bill with bluish-grey band, black breast and undertail-coverts, silvery-grey back, flanks and belly. Female has brown head with pale spot at the base of bill, white eye-ring, dark bill with bluish-grey band; greyish-brown back, flanks and belly. In flight shows greyish wing-bars. Eclipse male is similar to breeding male but duller.

Similar species Canvasback is larger, with much longer dark bill. Male Redhead has darker back and flanks; female Redhead is browner with bluish-grey bill with black tip.

Tadao Shimba

Adult ♂ breeding. February, Nagano

Tadao Shimba

Adult ♀. Feburary, Nagano

Tadao Shimba

Adult ♂s. February, Nagano

Range Breeds in northern Eurasia from Europe east to Lake Baikal. Moves south in winter.

Status in Japan Common winter visitor to ponds, lakes and coastal shallow waters from Kyushu northward. Common in city parks. A few breed in Hokkaido.

Canvasback

Aythya valisineria 55cm

Kazuyasu Kisaichi

Adult ♂ breeding. January, Ibaragi

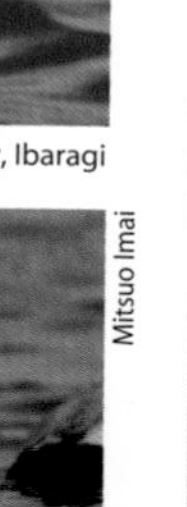

Mitsuo Imai

Adult ♀. February, Ibaragi

Description Large diving duck. Breeding male similar to male Common Pochard but larger, and has long darker bill and longer neck, back whiter. Female is similar to female Common Pochard but larger and paler, with longer black bill. Eclipse male similar to breeding male but browner.

Similar species Common Pochard is smaller, has shorter neck and shorter bill with blue-grey band.

Range Breeds in North America from Alaska south to northern prairie regions. Winters in southern North America.

Status in Japan Rare winter visitor to lakes, rivers and shallow coastal waters in northern Japan.

Redhead

Aythya americana 46-56cm

Tokio Sugiyama

Adult ♂ breeding. March, Vancouver, Canada

Description Breeding male has brick-red head, black breast and grey back and flanks; pale-blue bill separated by narrow white ring from black tip. Female is brown overall, with buff-coloured patch at base of bill. Eclipse male similar to breeding male but browner.

Similar species Male Common Pochard has paler back and flanks and dark bill with grey band; female Common Pochard is greyer.

Range Breeds in freshwater marshes in north-west North America and winters in southern United States.

Status in Japan Vagrant. One record of female in January 1985 in Tokyo.

Ring-necked Duck

Aythya collaris 40cm

Description Medium-sized diving duck. Breeding male has glossy purplish-black peaked head, dark-grey bill with white sub-terminal band and black tip; white band at the base of bill, black back and breast, and grey flanks. Eyes yellow. Female is dark-brown with white eye-ring and eye-stripe behind eye; white patch from base of bill to chin, dark bill with white band. Eyes dark. In flight shows grey wing-bars. Eclipse male is browner below; eyes yellow.
Similar species Female Tufted Duck has small head-tufts and lacks eye-ring. Female Common Pochard has rounder head and greyish-brown back.

Range Breeds in northern North America, winters along both coasts of North America.
Status in Japan Rare winter visitor to lakes and ponds. Most records from ponds in city parks.

Mitsuo Imai

Adult ♂ breeding. January, Kanagawa

Yoshihito Goto

Adult ♀. January, Osaka

Ferruginous Duck

Aythya nyroca 41cm

Description Medium-sized diving duck. Breeding male has reddish-brown head, breast and flanks; white eyes, dark-grey bill with bluish-grey subterminal band and black tip, blackish-brown back and white undertail-coverts. Female is duller with dark eyes. Eclipse male resembles female but eyes white.
Similar species Baer's Pochard is slightly larger, has longer bill and white on flanks. Male has glossy greenish-black head, female has pale brown loral spot.

Kazuyasu Kisaichi

Adult ♂ breeding. February, Fukuoka

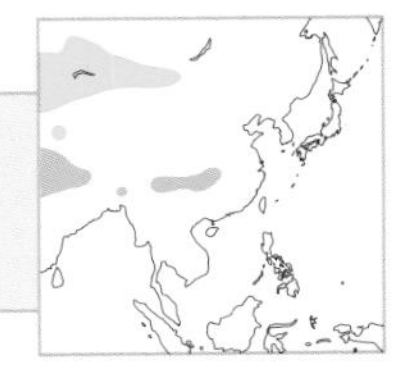

Range Breeds in temperate Eurasia from central and southern Europe east through Caspian Sea to western China. Winters in southern Europe, northern Africa and southern Asia.
Status in Japan Rare winter visitor to lakes and ponds.

Baer's Pochard

Aythya baeri 45cm

Tadao Shimba

Adult ♂ breeding. March, Tokyo

Tadao Shimba

Adult ♂ breeding. March, Tokyo

Description Medium-sized diving duck. Breeding male has glossy greenish-black head, white eyes and bluish-grey bill with black tip. Reddish-brown breast and dark-brown back, brown flanks with white patch near breast, white belly and undertail-coverts. Female has dark-brown head, dark eyes, bluish-grey bill with black tip and pale brown loral spot at the base. Brown breast, back and flanks, white belly and undertail-coverts. In flight shows white wing-bars. Eclipse male resembles female but eyes white.

Similar species Ferruginous Duck is slightly smaller, has smaller bill and head, more reddish head and breast, and lacks white on flanks.

Range Breeds in far eastern Asia; Amur Basin, Ussuriland and northern China. Winters mainly in eastern China and rare winter visitor to Korea.

Status in Japan Rare winter visitor to lakes and ponds from Kyushu northwards. Often observed on ponds in city parks among flocks of diving ducks.

Tadao Shimba

Adult ♀. December, Fukuoka

Tufted Duck

Aythya fuligula 40cm

Description Medium-sized diving duck. Breeding male has purplish-black head, long crest, yellow eye and bluish-grey bill with black tip; black breast, back and undertail-coverts sharply contrast with white flanks. Female has dark-brown head with short crest, yellow eyes, dark-grey bill with bluish-grey band and black tip. Dark-brown breast and back and pale flanks. Some females have white at the base of bill and pale undertail-coverts. In flight shows white wing-bars. Eclipse male duller than breeding male with brown flanks.
Similar species Female Ring-necked Duck has white eye-ring and eye-stripe; also lacks crest. Female Lesser Scaup has peaked head and smaller black tip to bill.

Tadao Shimba

Adult ♂ breeding. December, Tokyo

Tadao Shimba

First-winter ♂. December, Tokyo

Range Breeds across northern Eurasia from Europe east to Kamchatka Peninsula, south to Transbaikalia, Sakhalin and Kurile Islands. Moves south in winter and common winter visitor to Korea.
Status in Japan Common winter visitor to lakes, ponds and rivers; common in city parks in winter and often tame. Some breed in Hokkaido.

Tadao Shimba

Adult ♀. December, Tokyo

Greater Scaup

Aythya marila 45cm

Tadao Shimba

Adult ♂ breeding. April, Vancouver, Canada

Description Medium-sized diving duck. Breeding male has green-glossed black head, yellow eyes and bluish-grey bill with black tip, black breast, tail and undertail-coverts, white back finely barred black and white flanks. Female is dark-brown with white patch at the base of bill. In flight shows white wing-bars. Eclipse male is similar to breeding male but duller.
Similar species Lesser Scaup is slightly smaller, has peaked head, smaller black tip to bill and has white wing-bars confined to secondaries.

Tadao Shimba

Adult ♂ breeding. April, Vancouver, Canada

Range Breeds on tundra in northern Eurasia east to Kamchatka Peninsula. Winters south to temperate regions of Eurasia; also in North America.
Status in Japan Common winter visitor to shallow coastal waters and large lakes. Often forms large flocks.

Tadao Shimba

Adult ♀. April, Vancouver, Canada

Lesser Scaup

Aythya affinis 38-46cm

Description Breeding male has purplish-black glossy peaked head, yellow eyes, bluish-grey bill with small black tip, black breast, tail and uppertail-coverts, white back finely barred with black, and white flanks. Female is dark-brown overall, yellow eyes, bluish-grey bill with small black tip and white patch at the base. In flight shows white on secondaries only. Eclipse male similar to breeding but duller.
Similar species Greater Scaup is larger, has round-shaped head, larger black tip to bill and longer white wingbars.

Range Breeds in North America from Alaska south to northern prairie regions. Winters in southern North America.
Status in Japan Rare winter visitor to lakes and ponds.

Kazuyasu Kisaichi

Adult ♂ breeding. January, Tokyo

Kazuyasu Kisaichi

Adult ♀. April, Tokyo

Spectacled Eider

Somateria fischeri 53cm

Description Breeding male unmistakable; green head with large white eye patches, white back and black belly. Female has brown head with pale 'spectacle' pattern; brown body with dark bars on back and flanks. Eclipse male dark grey with paler 'spectacles' and white wing-coverts.
Similar species Female separated from other eiders by pale 'spectacle' pattern on face.

Range Breeds on the arctic coasts of north-east Siberia and Alaska. Winters on the ocean south to the Bering Sea.
Status in Japan No records.

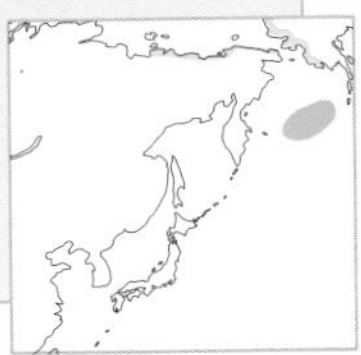

Pete Morris

Adults. June, Barrow, Alaska

Common Eider

Somateria mollissima 58–69cm

Tomi Muukonen

Adult ♂ breeding. May, Lemland, Finland

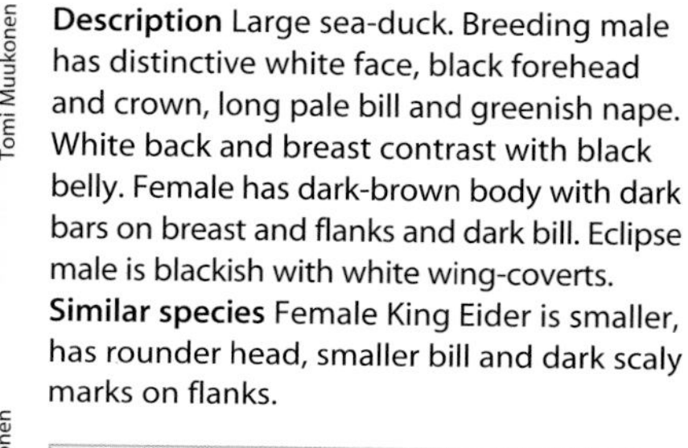

Description Large sea-duck. Breeding male has distinctive white face, black forehead and crown, long pale bill and greenish nape. White back and breast contrast with black belly. Female has dark-brown body with dark bars on breast and flanks and dark bill. Eclipse male is blackish with white wing-coverts.
Similar species Female King Eider is smaller, has rounder head, smaller bill and dark scaly marks on flanks.

Tomi Muukonen

Adult ♀. May, Lemland, Finland

Range Breeds in northern Eurasia from Europe east to Okhotsk coast; also in North America. Winters in northern seas.
Status in Japan Vagrant. One record: a female in 1971 in eastern Hokkaido.

King Eider

Somateria spectabilis 56cm

Tomi Muukonen

Adult ♂. March, Varanger, Norway

Description Breeding male unmistakable; red bill, black-bordered orange frontal shield, bluish-grey crown and nape. Female has dark-brown body with dark, scaly marks on flanks and small black bill. Eclipse male is blackish with red bill.
Similar species Female Common Eider is larger, has sloping head, larger bill and dark barred flanks.

Range Circumpolar. Breeds on coasts and islands of the Arctic Ocean and winters in northern seas. Winter range in north-eastern Asia south to Kuriles.
Status in Japan Vagrant. A few records from eastern Hokkaido. Occurs in coastal waters.

Steller's Eider

Polysticta stelleri 46cm

Description Breeding male unmistakable. White head with distinctive dark eye-patch, green spots on forehead and back of head; black chin, neck and back with white forewings, pale orange-brown breast and belly, and black tail and undertail-coverts. In flight shows white forewing and blue speculum with white trailing edge. Female is dark-brown with pale eye-ring and two broad, white wing-bars. In flight, shows blue speculum edged by white. Eclipse male is as female but forewings white.
Similar species Both female Common Eider and King Eider are larger with heavier bills; also lack white wing-bars.

Kazuyasu Kisaichi

Adult ♀. March, Norway

Range Breeds on lakes and tundra in northern Eurasia from Khatanga Bay east to Chukotski Peninsula; also in Alaska. Winters on the ocean in northern Europe, Alaska and in north-eastern Asia from Kamchatka south to Kurile Islands.
Status in Japan Rare winter visitor to coastal waters around eastern Hokkaido.

Tomi Muukonen

Adult ♀. March, Norway

Kazuyasu Kisaichi

Adult ♂. March, Norway

Harlequin Duck

Histrionicus histrionicus 43cm

Norio Yamagata

Adult ♂. March, Hokkaido

Description Male unmistakable. Dark-blue head, back and breast with white stripes on forehead, neck, breast and shoulders; chestnut flanks and small bluish-grey bill. Female is dark-brown, with white patch at the base of dark-grey bill, white spot on ear-coverts. In flight, wings all-dark.

Similar species Both female White-winged Scoter and Surf Scoter are larger, and have longer bills.

Norio Yamagata

Adult ♀. February, Hokkaido

Range Breeds in eastern Siberia from the Lena east to Kamchatka, south to Lake Baikal and Sakhalin; also in North America. Winters on rocky coasts from Okhotsk coast to Korea and Japan.

Status in Japan Uncommon winter visitor to rocky coasts from central Honshu northward. Some breed on mountains streams in northern Honshu.

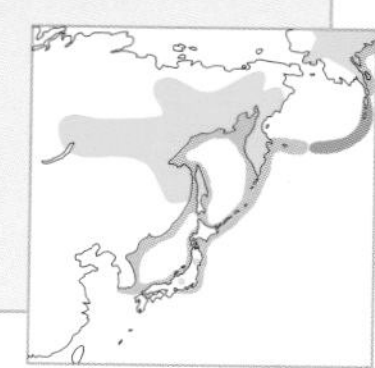

Mitsuo Imai

Adult ♀ (l) and ♂ (r). January, Hokkaido

Long-tailed Duck

Clangula hyemalis ♂60cm ♀38cm

Description Male in winter has white head, greyish face with dark patch on ear-coverts, white eye-ring and dark bill with pinkish band; black and white body with long black central feathers. In breeding plumage mainly chocolate-black, with white face and underparts and buff-coloured edges to back feathers. Female has white face with dark crown, dark-brown patch on face and dark bill; dark-brown breast and back with buff-coloured edges; white flanks. In breeding plumage female is dark-brown with grey face.

Teruaki Ishii

Adult ♂ (l) and ♀ (r). January, Hokkaido

Range Breeds on tundra in northern Eurasia from Scandinavia east to Chukotski Peninsula and across North America. Winters in coastal areas in northern Europe and in north-eastern Asia from Kamchatka Peninsula south to Korea and northern Japan.

Status in Japan
Common winter visitor to coastal waters on Hokkaido; uncommon on northern Honshu, rare elsewhere.

Teruaki Ishii

Adult ♀. February, Hokkaido

Teruaki Ishii

Adult ♂. February, Hokkaido

Black Scoter

Melanitta americana 48cm

Akira Yamamoto

Adult ♂. December, Aichi

Description Male black with distinctive yellow knob at the baseof bill. Female is dark-brown with pale cheeks. No white markings on wings in flight.
Similar species Both White-winged and Surf Scoters are larger, and have longer bills. Female Red-crested Pochard is browner and has white wing-bars.
Note Formely considered conspecific with Common Scoter *M. nigra*.

Kazuyasu Kisaichi

Adult ♀. March, Hokkaido

Range Breeds on lakes and tundra in northern Eurasia from the Yana east to Kamchatka; also in North America. Common winter visitor to coastal waters south to Korea.
Status in Japan Common winter visitor to coastal waters from Honshu northward. Some may breed in Hokkaido.

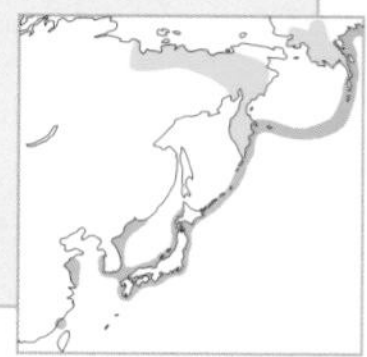

Akira Yamamoto

Adult ♂. December, Aichi

White-winged Scoter

Melanitta deglandi 55cm

Description Male is black with white crescent-shaped patch below eye, orange bill with black knob at base. White secondaries. Female is dark-brown, black bill with white spot at the base and on cheeks. In flight, shows white secondaries that are also visible when swimming.
Similar species Female Surf Scoter has white spot on hindneck and all dark secondaries.

Akira Yamamoto

Adult ♂. January, Aichi

Kazuyasu Kisaichi

Adult ♀. December, Hokkaido

Range Breeds from the Yenisey east to Kamchatka, south to Sakhalin and northern Kurile Islands; also in North America. Winters in temperate coastal waters.
Status in Japan Common winter visitor to coastal waters from Kyushu northward.

Surf Scoter

Melanitta perspicillata 56cm

Description Male black with distinctive white patches on forehead and hindneck, and large white, orange and black patterned bill with yellow tip. Female Dark-brown overall with dark bill and distinctive white patches at the base of bill, cheeks and hindneck. In flight, no white markings on wings.
Similar species White-winged Scoter has white secondaries.

Range Breeds on northern lakes in North America. Winters on the ocean and on shallow coastal waters.
Status in Japan Very rare winter visitor to coastal waters from central Honshu northwards.

Tokio Sugiyama

Adult ♂. March, Vancouver, Canada

Common Goldeneye

Bucephala clangula 45cm

Tadao Shimba

Adult ♂ breeding. January, Hokkaido

Description Medium-sized diving duck. Breeding male has glossy greenish-black head, yellow bill with diagnostic white patch at the base, white underparts with black scapulars and black tail–coverts. Female has dark-brown head with yellowish-white eye, dark bill with orange tip, white neck, dark brown back and greyish-brown breast and flanks. In flight, white wing–coverts and secondaries. Eclipse male as female but more extensive white on forewings. **Similar species** Only possible confusion is Barrow's Goldeneye *B. islandica;* has not been recorded in Japan.

Range Breeds in northern Eurasia from Scandinavia east to Kamchatka, south to Amur, Ussuriland and Sakhalin; also in North America. Winters to temperate regions and common winter visitor to Korea.
Status in Japan Common winter visitor to shallow waters from Kyushu north.

Bufflehead

Bucephala albeola 35.5cm

Tadao Shimba

Adult ♂ breeding. March, Ontario, Canada

Description Small diving duck. Breeding male has glossy purplish-black head with distinctive large white patch from behind eye to back of head. Bluish-grey bill, dark back, white breast and flanks and dark-grey tail. In flight, white wing-coverts and secondaries contrast with dark primaries. Female has dark-brown head with oval-shaped white patch on ear-coverts, dark bill, dark back, and greyish-brown breast and flanks. In flight shows white secondaries on upperwings. Eclipse male is duller than breeding male and grey below.

Range Breeds on wooded lakes in northern North America and winters in southern United States.
Status in Japan Rare winter visitor to shallow coastal waters in northern Japan.

Smew

Mergellus albellus 42cm

Description Breeding male is white with distinctive black eye-patch from the base of bill. White crest with black patch and dark-grey bill, two black lines on breast, black back and shoulder-line; white flanks, finely barred grey. In flight, white upperwing-coverts and dark secondaries with white tips. Female has chestnut crown and hindneck, black at base of bill, white chin and upper-neck, dark greyish breast, back and flanks. In flight, white upperwing-coverts and dark secondaries with white tips. Eclipse male resembles female but has darker back and lores.
Similar species Both Slavonian and Black-necked Grebes in non-breeding plumage have darker caps, longer necks and red eyes.

Tokio Sugiyama

Adult ♂ breeding. January, Aichi

Tokio Sugiyama

Adult ♀. January, Aichi

Range Breeds on wooded lakes and river banks in northern Eurasia from Scandinavia east through Siberia to Kamchatka Peninsula and Sakhalin. Moves south in winter.
Status in Japan Uncommon winter visitor to lakes, freshwater marshes and rivers. A few breed in Hokkaido.

Nobuo Shiraki

Adult ♂ breeding. February, Osaka

Red-breasted Merganser

Mergus serrator 55cm

Akira Yamamoto

Adult ♂ breeding. April, Aichi

Akira Yamamoto

Adult ♀. April, Aichi

Description Breeding male has glossy greenish-black head with shaggy crest, red eyes and bill, white collar; brown breast with fine dark streaks, black back with white shoulder patch and finely barred grey flanks. In flight, white wing-coverts and secondaries with black tips to median and greater coverts. Female has chestnut-brown head with shaggy short crest, red eyes and bill, white throat and foreneck, greyish-brown back and flanks. In flight shows grey forewings, black primaries and white secondaries with black tips to greater coverts. Eclipse male resembles female but retains white upperwing-coverts.
Similar species Female Goosander is larger and has sharper contrast between chestnut upper neck and greyish lower neck.

Range Breeds in northern Eurasia from northern Europe east through Siberia to Kamchatka Peninsula, south to Amur Basin, Ussuriland, Sakhalin and north-eastern China; also in North America. Winters to the temperate regions and common winter visitor to Korea.
Status in Japan Common winter visitor to coastal waters; rarely occurs inland.

Goosander

Mergus merganser 65cm

Other name Common Merganser
Description Breeding male has glossy greenish-black head, red bill, dark neck and black back; white breast, belly and flanks, and grey uppertail-coverts and tail. In flight shows white upperwing-coverts and secondaries contrast with black primaries. Female has chestnut head with short shaggy crest, grey back and flanks. Chestnut upper neck and greyish breast clearly separated by white collar. In flight shows grey upperwing-coverts and white secondaries. Eclipse male resembles female but has white upperwing-coverts.
Similar species Female Red-breasted Merganser is smaller and lacks white collar dividing chestnut head from greyish breast.

Tadao Shimba

Adult ♂ breeding. February, Tokyo

Tokio Sugiyama

Adult ♀. February, Gifu

Range Breeds in northern Eurasia from northern Europe east through Siberia to Kamchatka Peninsula, south to Amur Basin, Ussuriland, Sakhalin and north-east China; also in North America. Winters south to temperate regions and common winter visitor to Korea.
Status in Japan Common winter visitor to lakes, rivers and coastal waters from Kyushu northwards. Some breed in Hokkaido.

Scaly-sided Merganser

Mergus squamatus 52–58cm

Norio Yamagata

Adult ♂ breeding. December, Gifu

Norio Yamagata

Adult ♀. April, Shiga

Yoshihito Goto

Adult ♂ breeding. February, Gifu

Description Breeding male has glossy greenish-black head with very long shaggy crest, red bill, black neck and back; white breast, belly and conspicuous dark scaly marks on flanks. In flight shows white upperwing-coverts and secondaries, with black tips to median and greater coverts. Female Chestnut head with short shaggy crest, dark lores and red bill; white breast, grey back and conspicuous dark scaly marks on flanks. In flight shows white secondaries with black tips to greater coverts. Eclipse male resembles female but has white upperwing-coverts.

Similar species Red-breasted Merganser lacks scaly marks on flanks.

Range Breeds in Ussuriland, Amur Basin and north-eastern China and winters mainly in eastern and southern China; rare winter visitor to Korea.

Status in Japan First recorded in February 1986 in Gifu and regular in western Japan since then. Rare winter visitor to lakes and rivers.

Osprey

Pandion haliaetus ♂54cm ♀64cm WS155–175cm

Description Graceful, long-winged fishing hawk with broad dark eye-stripe and dark-brown back, white underparts with dark carpal patches and dark flight feathers. Female has broad breast-band. Immature has buff-coloured edges to upperpart feathers.
Similar species Separated from other raptors by long, narrow wings and white underparts.

Akira Yamamoto

Adult. February, Gifu

Akira Yamamoto

Juvenile. January, Aichi

Range Cosmopolitan. Northern populations are migratory and moves south in winter. In north-eastern Asia, breeds from Kamchatka Peninsula, Okhotsk coast south to north-eastern China; mainly a summer visitor.
Status in Japan Common resident in coastal areas, lakes and large rivers. Summer visitor to Hokkaido and winter visitor to Ryukyu Islands.

Akira Yamamoto

Adult. February, Aichi

Oriental Honey-buzzard

Pernis ptilorhynchus ♂57cm ♀61cm WS121–135cm

Tokio Sugiyama
Adult ♂. July, Aichi

Description Long-necked hawk with long wings. Male has greyish head, dark-brown back, white throat and dark-brown foreneck. Underparts variable, usually brownish streaked darker. Dark bands across flight feathers, pale tail with bold, dark bands. Female browner. Juvenile has four or five narrow tail-bands. Generally brown, head and neck paler, sometimes whitish.
Similar species Mountain Hawk Eagle is larger, has shorter neck and broader wings.

Nobuo Shiraki
Adult ♂. September, Nagasaki

Range Breeds from eastern Siberia east to Sakhalin; south to north-east China and Japan. Winters in south-east Asia. Uncommon passage migrant to Korea.
Status in Japan Common summer visitor to wooded areas from Kyushu northwards, feeding mainly on larvae at nests of wasps and bees. Highly migratory and flocks of hundreds can be observed at raptor migration vantage points in spring and autumn.

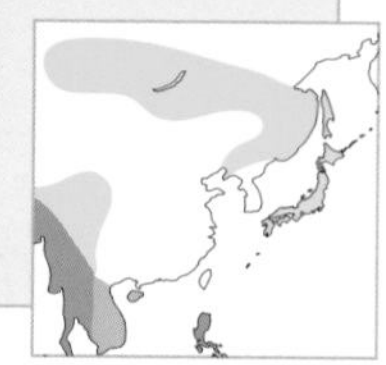

Nobuo Shiraki
Adult ♀. September, Nagasaki

Shuichi Tsuno
Juvenile. September, Wakayama

Shuichi Tsuno
Juvenile. October, Wakayama.

Black Kite

Milvus migrans ♂58.5cm ♀68.5cm WS157–162cm

Description Large, dark-coloured hawk with long, angled wings and shallow forked tail. Dark-brown overall with pale face and black eye-patch, pale wing-coverts on upperwings and distinctive white flash at the base of primaries on underwing. Immature shows buff-coloured streaks below and pale feather tips above. In flight, slow wingbeats with long glides. Frequently circles high in the sky. **Similar species** Separated from other raptors by distinctive white patches at the base of primaries and shallow forked tail.

Range Widely distributed across Eurasia, Africa and Australia. **Status in Japan** Most common hawk in Japan. Flocks of hundreds often congregate at fishing harbours to scavenge discards and waste.

Tadao Shimba

Adult. November, Aichi.

Tadao Shimba

Juvenile. December, Hokkaido

Tadao Shimba

Adults. February, Nagano

White-tailed Eagle

Haliaeetus albicilla ♂80cm ♀95cm WS182–230cm

Manabu Totsuka

Adults. February, Hokkaido

Description Large, dark-brown sea-eagle with distinctive white wedge-shaped tail. Adult has pale-brown head, large yellow hooked bill and yellow legs. Juvenile has dark-brown head and white-mottled dark underparts and tail, dark eye and dark bill. Adult plumage is attained after several years.
Similar species Immature Steller's Sea Eagle is larger, has broader wings, longer tail and larger, yellow bill.

Range Widely distributed across northern Eurasia.
Status in Japan Locally common winter visitor to northern Japan; rare elsewhere. Inhabits coastal areas, lakes and rivers. Small numbers breed in eastern and northern Hokkaido. Large flocks can be seen in eastern Hokkaido in winter. Feeds mainly on fish, ducks and carrion.

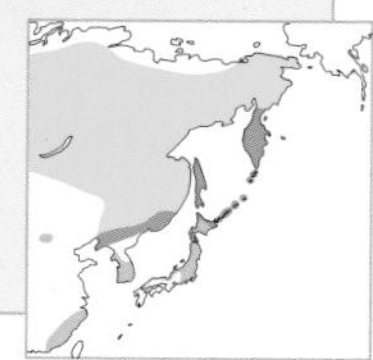

Tadao Shimba

Adult. December, Hokkaido

Tadao Shimba

Immature. December, Hokkaido

Steller's Sea Eagle

Haliaeetus pelagicus ♂88cm ♀102cm WS220–245cm

Description Massive sea-eagle with very large yellow bill. In adult, dark-brown body contrasts with white forewings, long, white wedge-shaped tail and white thighs. Juvenile has white-mottled dark-brown wings and tail, yellowish facial skin and yellow bill. Adult plumage is attained after several years.
Similar species Juvenile White-tailed Eagle has smaller dark bill, shorter tail and narrower wings.

Range Breeds from Kamchatka and shores of Sea of Okhotsk south to Sakhalin. In winter, moves south to Korea and northern Japan.
Status in Japan Locally common winter visitor to northern Japan; rare elsewhere. Inhabits coastal areas, lakes and rivers. Large flocks can be seen in eastern Hokkaido in winter. Feeds mainly on fish, ducks and carrion.

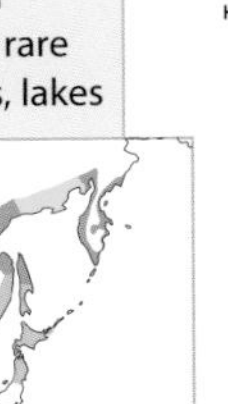

Manabu Totsuka

Adult. February, Hokkaido

Tadao Shimba

Immature. December, Hokkaido

Manabu Totsuka

Adult. February, Hokkaido

Eurasian Black Vulture

Aegypius monachus 100–110cm WS250–295cm

Hyun-tae Kim

Immature. December, Sosan, South Korea

Other name Cinereous Vulture
Description Largest raptor in the region. Dark-brown body, very long, broad wings and short wedge-shaped tail. Adult has bare skin on head. Immature has head feathered blackish-brown and darker body.

Nobuo Shiraki

Immature. June, Mongolia

Range Southern Europe east through central Asia to Mongolia and north-east China. Asian population is migratory and locally common winter visitor to Korea.
Status in Japan Very rare winter visitor to cultivated fields, wooded areas and coastal areas. Feeds on carrion. Birds in Japan often arrive exhausted and remain until they have recovered.

Nobuo Shiraki

Immature. June, Mongolia

Crested Serpent Eagle

Spilornis cheela 53–55cm WS120cm

Description Small eagle of subtropical forests. Black crown; white spotted, dark-brown back and pale brown underparts with white spots; yellow eyes and lores. In flight, diagnostic white band across flight feathers and broad white tail-band visible. Soars with wings slightly uplifted. Juvenile has mottled white and brown back, white face and underparts.
Voice Shrill, screaming disyllabic *kwee-kwee* or loud, monotonous and repeated *pyot-pyot-pyot*.
Similar species Oriental Honey-buzzard is larger with longer neck and narrowly-barred wings.

Range Resident in India, south-east Asia, and southern China east to Taiwan.
Status in Japan Race *perplexus* locally common resident on Yaeyama Islands; this is often considered a separate species, Ryukyu Serpent Eagle. Hunts from perch, preferring forest edges around rice fields and freshwater marshes.

Akira Yamamoto

Adult. January, Iriomote-jima

Akira Yamamoto

Immature. January, Iriomote-jima

Nobuo Shiraki

Adult. January, Ishigaki-jima

Eastern Marsh Harrier

Circus spilonotus ♂48cm ♀56cm WS113–137cm

Norio Yamagata
Adult ♂. January, Yamaguchi

Nobuo Shiraki
Adult ♂. December, Kagoshima

Tsunehiro Komai
Adult ♂. March, Aichi

Description Long-winged hawk that glides over marshes with raised wings. Male plumage is variable. Resident form has finely streaked brown head and back, whitish uppertail-coverts and barred brown tail. In flight, brown wing-coverts contrast with grey flight feathers and black primary tips. Some males visiting Japan in winter from north-east China or far-eastern Russia have distinctive black heads, black upperwing-coverts and white underparts with dark streaked breast. Females have streaked brown body and wings with dark primary tips and barred tail. Juvenile is similar to female, but has cream-coloured head.

Norio Yamagata
Adult ♀. February, Kagoshima

Similar species Male Pied Harrier is similar to black-headed form of male, but has clear anchor-shaped black pattern on upperwings. Female and juvenile Pied and Hen Harriers are smaller and narrower-winged.

Range Breeds from Mongolia east through north-east China to Sakhalin and winters in south-eastern Asia.
Status in Japan Common winter visitor to large reed beds and grassy fields. Breeds locally from central Honshu northwards.

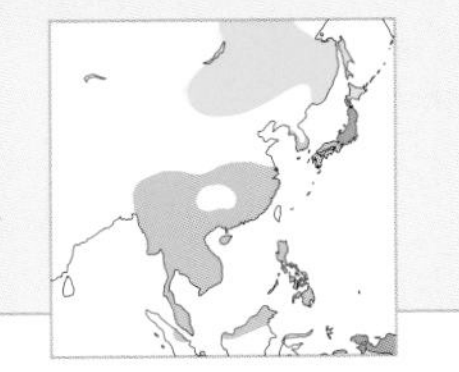

Akira Yamamoto

Immature. December, Aichi

Nobuo Shiraki

Adult ♂. February, Aichi

Tsunehiro Komai

Adult ♂. December, Aichi

Hen Harrier

Circus cyaneus ♂43cm ♀53cm WS98.5–123.5cm

Akira Yamamoto

Adult ♀. February, Nara

Akira Yamamoto

Adult ♀. February, Nara

Description Slender long-winged hawk. Male has grey head, breast, back and tail with clear white uppertail-coverts. In flight, white belly and underwings contrast with black primaries. Female has dark greyish-brown upperparts and clear white uppertail-coverts, yellowish-white underparts with heavy brown streaks. Juvenile resembles female but cheek crescents darker.

Similar species Male Eastern Marsh Harrier has brown or black head and upperwing coverts. Female Pied Harrier has bluish-grey upper flight feathers and anchor-shaped brown pattern on upperwings.

Range Breeds across northern Eurasia and northern North America and moves south in winter.

Status in Japan Uncommon winter visitor to cultivated and grassy fields.

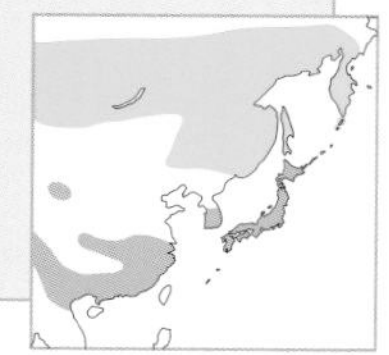

Shuichi Tsuno

Adult ♂. December, Aichi

Pied Harrier

Circus melanoleucus ♂41–44cm ♀45cm WS104–115cm

Description Male has distinctive black head and breast, bluish-grey upperparts with anchor-shaped black pattern on upperwings and black primaries, bluish-grey tail and white uppertail-coverts. Female has dark-brown head, dark-brown anchor-shaped pattern on upperwings and dark-brown bars across bluish-grey flight feathers above, bluish-grey tail with dark-brown bands and white uppertail-coverts.
Similar species Female Hen Harrier has more uniform dark-brown upperwings.

Norio Yamagata

Adult ♂. August, Tochigi

Range Breeds mainly in Ussuriland and north-eastern China and winters in south-east Asia. Rare passage migrant to Korea.
Status in Japan Rare passage migrant to fresh-water marshes and rice fields. Attempted nesting in 1989 in Aichi.

Akira Yamamoto

Adult ♀. June, Aichi

Akira Yamamoto

Adult ♀. June, Aichi

Chinese Goshawk

Accipiter soloensis ♂30cm ♀33cm WS53–64cm

Akira Yamamoto

Adult ♂. September, Miyako-jima

Nobuo Shiraki

Adult ♀. September, Nagasaki

Description Small, highly migratory hawk. In flight, black primary tips contrast with white underwings; four 'fingered' primaries and unbarred underwing-coverts. Adult male has bluish-grey head and dark bluish-grey back, white underparts with pale orange-brown breast. Female has dark bluish-grey head and back, white underparts with orange-brown breast. Juvenile has greyish-brown back, whitish underparts with heavy streaks on breast and boldly barred belly. **Similar species** Japanese Sparrowhawk has dark stripe in centre of white throat, shorter wings and barred underwing-coverts; also five 'fingered' primaries visible in flight.

Range Restricted to far-eastern Eurasia. Breeds from southern Ussuriland south to Korea and winters in south-east Asia. **Status in Japan** Locally abundant passage migrant to Kyushu and Ryukyu Islands. Flocks of thousands can be observed on good migration days in autumn.

Nobuo Shiraki

Juvenile. September, Nagasaki

Japanese Sparrowhawk

Accipiter gularis ♂27cm ♀30cm WS51–63cm

Description Smallest hawk in Japan. Five 'fingered' primaries visible in flight. Adult male has dark bluish-grey head and back with pale rufous breast and flanks. Female is larger than male, with brownish-grey head and back, indistinct supercilium and heavily barred underparts. Juvenile has dark-brown back, heavily streaked belly and single dark stripe on centre of white throat.
Similar species Chinese Goshawk has longer wings with dark primary tips, unbarred underwing-coverts and four 'fingered' primaries visible in flight.

Tadao Shimba

Adult ♀. November, Aichi

Range Breeds from northern Mongolia east to Sakhalin, south to Korea and Japan. Winters in southern China and south-eastern Asia.
Status in Japan Common summer visitor to wooded areas from central Honshu northwards. A few breed in urban wooded areas. Common passage migrant elsewhere and some winter in southern Japan. Hundreds can be observed on good migration days in autumn.

Tadao Shimba

Juvenile. November, Aichi

Kazuyasu Kisaichi

Adult ♂. June, Chiba

Tadao Shimba

Juvenile. October, Aichi

Eurasian Sparrowhawk

Accipiter nisus ♂32cm ♀39cm WS62–76cm

Tokio Sugiyama

Adult ♀. January, Hokkaido

Tokio Sugiyama

Adult ♀. January, Hokkaido

Description Pigeon-sized hawk with short wings and long tail. Six 'fingered' primaries visible in flight. Adult male has bluish-grey back, indistinct supercilium and whitish underparts with rufous-brown breast and belly. Female is larger than male with white supercilium and finely barred underparts. Immature has greyish-brown back, white underparts with heavy dark barring. Preys almost exclusively on small birds.

Similar species Japanese Sparrowhawk has dark stripe in centre of throat, shorter tail and five 'fingered' primaries visible in flight.

Range Widely distributed across Eurasia, northern populations move south in winter.

Status in Japan Common resident, breeding in mountain forests in Honshu and open forests in Hokkaido. Winters in wooded areas and cultivated fields.

Nobuo Shiraki

Juvenile. April, Wakayama

Northern Goshawk

Accipiter gentilis ♂50cm ♀58.5cm WS105–130cm

Description Crow-sized hawk with short, broad wings, with 'bulging' secondaries and long, rounded tail. Adult has conspicuous broad white supercilium, dark bluish-grey back and finely barred white underparts. Female is larger than male with more heavily barred underparts. Immature has brown back with buff-coloured feather edges, pale yellowish-brown underparts with heavy dark brown streaks. **Similar species** Eurasian Sparrowhawk is smaller, has narrower wings and square-tipped tail.

Range Northern Eurasia and northern North America. Partially migratory.
Status in Japan Common resident; breeds in wooded areas from Kyushu northward. Winters in wooded areas, cultivated fields and estuaries. Larger and paler eastern Siberian race *albidus* has occurred in northern Japan.

Nobuo Shiraki

Adult ♀. June, Wakayama

Nobuo Shiraki

Adult ♀. June, Wakayama

Tadao Shimba

Juvenile. November, Aichi

Grey-faced Buzzard

Butastur indicus ♂47cm ♀51cm WS105–115cm

Nobuo Shiraki

Adult ♂. June, Osaka

Description Slender, reddish-brown, highly migratory hawk. Greyish-brown head and white throat with central black stripe, reddish-brown back and breast. Underwings whitish, barred brown on coverts and grey on remiges. White belly heavily barred reddish-brown. Grey tail with three dark bands in adult, four to five in juvenile. Female is paler than male and has white supercilium.

Voice Call is a loud and distinctive disyllabic *pit-kwee*.

Similar species Oriental Honey-buzzard is larger and has longer slender neck.

Nobuo Shiraki

Adult ♀. June, Osaka

Range Breeds from Amur Basin south through north-eastern China to Japan. Winters mainly in southern China and south-east Asia. Uncommon passage migrant to Korea.

Status in Japan Common summer visitor to wooded areas from Honshu to Kyushu. Large flocks can be seen at raptor migration vantage points in autumn. Common winter visitor to the Ryukyu Islands.

Nobuo Shiraki

Adult ♀. April, Wakayama

Common Buzzard

Buteo buteo ♂50–53cm ♀53–60cm WS122–137cm

Description Crow-sized, broad-winged hawk. Brown head and back, pale buff-coloured underparts with brown band across belly. Broad wings with dark carpal-patches and black primary tips visible in flight. Immature has light brown back and yellow eyes. Adult has dark eyes.
Similar species Rough-legged Buzzard has white tail with conspicuous dark subterminal band. Upland Buzzard is larger, has distinctive large white panel above at the base of primaries and half-feathered tarsus.

Akira Yamamoto

Juvenile *japonicus*. January, Aichi

Range Widely distributed across temperate Eurasia, northern birds move south in winter.
Status in Japan Common resident. Breeds in mountain forests from Shikoku to Hokkaido. Winters in warmer areas and often observed around forest edges and cultivated fields. Race *toyoshimai* on Ogasawara Islands is paler overall.

Shuichi Tsuno

Adult *japonicus*. October, Wakayama

Atle Ivar Olsen

Adult *toyoshimai*. April, Haha-jima

Upland Buzzard

Buteo hemilasius ♂61cm ♀72cm WS158cm

Hyun-tae Kim

Adult, pale morph. January, Sosan, South Korea

Hyun-tae Kim

Adult, pale morph. January, Sosan, South Korea

Description Similar to Common Buzzard, but larger. Dimorphic; pale morph has a whitish head, pale-brownish back and dark-brown lower belly; tarsus is half-feathered. In flight shows whitish underwings with dark carpal-patches and dark primary tips, brown upperwings with distinctive large white panel at the base of primaries, pale-brownish tail with several indistinct bands. Dark morph blackish-brown with grey remiges; whitish tail with broad, dark subterminal band.
Similar species Rough-legged Buzzard is paler, has white tail with distinctive dark subterminal band. Common Buzzard is smaller, has uniform brown tail and unfeathered tarsus.

Range Breeds from Altai east to north-eastern China. Winters from central Asia east to Korea.
Status in Japan Rare winter visitor to cultivated and grassy fields.

Norio Yamagata

Juvenile. December, Kagoshima

Rough-legged Buzzard

Buteo lagopus ♂53–57cm ♀60cm WS129–143cm

Description Medium-sized raptor with distinctive black subterminal band to white tail. Mottled white and dark-brown back, white underparts with dark-brown breast and belly. In flight shows white underwings with black carpal-patches and black tips to primaries, also black trailing-edges to flight feathers. Immature has whitish head, white mottled dark-brown back and whitish breast contrasting with dark belly.

Similar species Common Buzzard has uniform brown tail. Upland Buzzard is larger, has a distinctive large white panel above at the base of primaries, and a pale-brownish tail with several indistinct bands.

Yoshihito Goto

Juvenile. December, Aichi

Yoshihito Goto

Adult ♂. February, Aichi

Range Northern Eurasia and northern North America. Northern birds move south in winter. Rare winter visitor to Korea.

Status in Japan Rare winter visitor to cultivated and grassy fields.

Tadao Shimba

Juvenile. December, Aichi

Greater Spotted Eagle

Aquila clanga ♂67cm ♀70cm WS158–182cm

Tokio Sugiyama

Adult. December, Kagoshima

Description Smaller than other dark-brown eagles. Yellow gape conspicuous. Adult has blackish-brown plumage with short dark tail and dark-brown upperparts. Juvenile has dark-brown head, dark-brown wings with conspicuous buff-coloured spots on scapulars and wing-coverts. In flight, white spots form pale lines across upperwings.

Similar species Eastern Imperial Eagle is larger, has longer tail, yellowish-brown crown and hindneck.

Norio Yamagata

Juvenile. January, Kagoshima

Range Breeds from Eastern Europe east to far-eastern Russia and north-east China. Moves south in winter.

Status in Japan Very rare winter visitor to cultivated fields and wooded areas. A single bird visited Kagoshima from 1992 until 2006.

Norio Yamagata

Adult. December, Kagoshima

Eastern Imperial Eagle

Aquila heliaca ♂77cm ♀83cm WS195–207cm

Description Large eagle. Adult is dark blackish-brown with yellowish-brown crown and hindneck; white patches on scapulars, grey uppertail with dark bars. Juvenile has buff-coloured head and buff-coloured streaks on brown body. In flight, buff-coloured forewings and pale inner primaries contrast with dark flight feathers. Soars with wings flat.

Similar species Adult Golden Eagle is browner, has golden-brown hindneck and back of head; soars with wings slightly uplifted and has bulging secondaries.

Hyun-tae Kim

Adult. January, Sosan, South Korea

Range Breeds from Eastern Europe through central Asia to Lake Baikal area. Moves south in winter.

Status in Japan Very rare winter visitor to cultivated fields, grassy fields and riverbanks. Recorded from Honshu southwards.

Hyun-tae Kim

Adult. January, Sosan, South Korea

Teruaki Ishii

Juvenile. January, Hong Kong

Golden Eagle

Aquila chrysaetos ♂78–86cm ♀85–95cm WS170–210cm

Akira Yamamoto

Adult. September, Shiga

Description Large, dark-brown eagle. Adult has golden wash over back of head and hindneck, and pale panel across upperwing-coverts. Immature has well-defined white patches at base of primaries and white patch at base of tail. Soars with wings slightly uplifted.
Similar species Immature White-tailed Eagle has shorter, wedge-shaped tail and broader wings.

Kazuyasu Kisaichi

Juvenile. November, Shiga

Range Widely distributed across Eurasia, northern Africa and North America.
Status in Japan Uncommon resident from Hokkaido to Kyushu. Nests on rocky ledges and in large trees in mountains. Pairs remain within breeding territory all year.

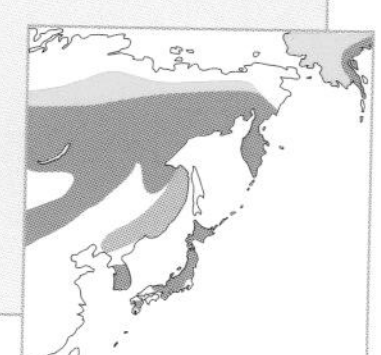

Tokio Sugiyama

Adult. August, Shiga

Mountain Hawk Eagle

Spizaetus nipalensis ♂70–75cm ♀77–83cm WS140–165cm

Description A large eagle with broad wings and a short, rounded tail. Adult has a dark-brown face and crown with crest, white throat with black central stripe, brown back and pale-brown underparts. In flight, finely barred pale-brown underwings and heavily barred tail visible. Juvenile has paler throat and underparts.

Similar species Oriental Honey-buzzard has longer slender neck and narrower wings.

Tokio Sugiyama

Adult. June, Aichi

Range Western India east to northern parts of south-east Asia, southern China and Taiwan.

Status in Japan Uncommon resident in mountain forests from Kyushu northwards.

Nobuo Shiraki

Juvenile. March, Shizuoka

Akira Yamamoto

Adult. July, Shiga

Lesser Kestrel

Falco naumanni ♂28cm ♀31cm WS61–66cm

Tomi Muukonen
Adult ♂. May, Greece

Tomi Muukonen
Adult ♂. May, Greece

Description Similar to Common Kestrel but slightly smaller and slimmer, with shorter, wedge-shaped tail. Adult male has bluish-grey head without moustaches; bright, unspotted chestnut back and bluish-grey tail with dark subterminal band, buff-coloured underparts with a few dark spots. Adult female has reddish-brown head and back with fine dark bars; underparts buff-coloured with dark streaks. Both sexes have white claws.

Range Breeds from southern Europe east through central Asia to north-western China. Winters in central and southern Africa. Often hunts for insects in large flocks.
Status in Japan Vagrant. A few records from Tsushima and Yaeyama Islands. Inhabits cultivated fields.

Nobuo Shiraki
Immature ♂. June, Mongolia

Common Kestrel

Falco tinnunculus ♂28cm ♀31cm WS61–66cm

Description Small falcon with long wings and tail. Adult male has bluish-grey head with dark moustache and dark spot behind eye; chestnut-brown back with dark spots and bluish-grey tail with black subterminal band, buff-coloured underparts with heavy dark streaks. Adult female has reddish-brown head, back heavily barred dark-brown. Immature is similar to female. Frequently hovers over open country.
Similar species Lesser Kestrel is slightly smaller, has shorter tail. Male Lesser has unmarked rufous back and female has less barred back; also have white (not dark) claws at close range.

Range Distributed widely across Eurasia and Africa. Northern populations are partly migratory.
Status Common resident; breeds in eastern Japan. Occurrence of nesting on man-made structures such as buildings and bridges is increasing. Common winter visitor elsewhere inhabiting cultivated and grassy fields and riverbanks.

Tokio Sugiyama

Adult ♂. February, Aichi

Tokio Sugiyama

Juvenile. February, Aichi

Tadao Shimba

Juvenile. December, Hokkaido

Amur Falcon

Falco amurensis ♂28cm ♀31cm WS69–76cm

Ran Schols

Adult ♂. May, Bedaihe, China

Description Adult male has slaty-grey head and back, white cheeks and throat. Dark bluish-grey underparts with chestnut lower belly and undertail-coverts. In flight, white underwing–coverts contrast with dark flight feathers. Adult female has dark bluish-grey back with black bars, dark moustache and whitish underparts, with dark streaked breast. Juvenile is similar to female, but has white supercilium and buff-coloured feather edges above.

Similar species Juvenile Northern Hobby has darker back and prominent broad moustache.

Mitsuo Imai

Adult ♀. May, Hegura-jima

Range Breeds in Ussuriland and north-east China and winters in southern Africa.

Status in Japan Rare passage migrant to cultivated fields. Often seen on telephone wires in cultivated land.

Mitsuo Imai

Adult ♀. May, Hegura-jima

Merlin

Falco columbarius ♂28cm ♀31cm WS61–66cm

Description Small falcon of open country. Adult male has bluish-grey crown and back, white supercilium and throat, orange-brown underparts, heavily streaked dark-brown, bluish-grey tail with dark subterminal band. Adult female has greyish-brown crown and back; cream-coloured underparts boldly marked with dark-brown streaks. Immature is browner.

Similar species Common Kestrel is larger, has reddish-brown back, longer tail and frequently hovers.

Tokio Sugiyama

Adult ♂. January, Aichi

Range Breeds across northern Eurasia and northern North America and moves south in winter.

Status in Japan Uncommon winter visitor to cultivated and grassy fields. Rare on Ryukyu Islands. Preys mainly on small birds.

Akira Yamamoto

Juvenile. January, Aichi

Akira Yamamoto

Juvenile. January, Aichi

Northern Hobby

Falco subbuteo ♂34cm ♀37cm WS72–84cm

Nobuo Shiraki
Adult. July, Hokkaido

Description Small, agile falcon with long pointed wings. Adult has dark-grey upperparts, heavily streaked belly with characteristic chestnut undertail-coverts. Immature has dark-brown upperparts, more heavily streaked belly and lacks chestnut on undertail-coverts. **Similar species** Peregrine Falcon is larger, with stouter body and broader wings.

Nobuo Shiraki
Adult. July, Hokkaido

Range Breeds across northern Eurasia east to Kamchatka Peninsula and Okhotsk coast. Moves south in winter. **Status in Japan** Uncommon summer visitor from northern Honshu northwards. Sometimes breeds in wooded urban areas. Uncommon passage migrant elsewhere, mainly in autumn.

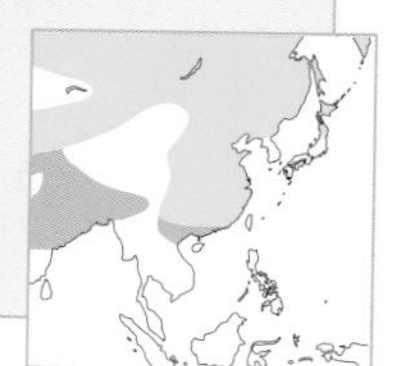

Shuichi Tsuno
Juvenile. October, Wakayama

Gyrfalcon

Falco rusticolus ♂56cm ♀61cm WS124–132cm

Description Large, heavily-built falcon with broad wings. Plumage varies from white to dark brownish-grey. White morph is mainly white, variably marked with black feather tips on wings, back and tail. Grey morph has dark bluish-grey back heavily barred with white, white face with dark crown, dark-stripe over eye and dark moustaches; underparts whitish with fine barring. Immature is browner.

Similar species Peregrine Falcon is smaller, has dark greyish back and prominent moustaches.

Yoshihito Goto

Adult white morph. December, Hokkaido

Range Circumpolar. Breeds in Arctic Eurasia, North America and Greenland. Some move south in winter.

Status in Japan Rare winter visitor, mainly to Hokkaido. Occurs in coastal areas and large open fields. Tends to fly low over ground in search of prey.

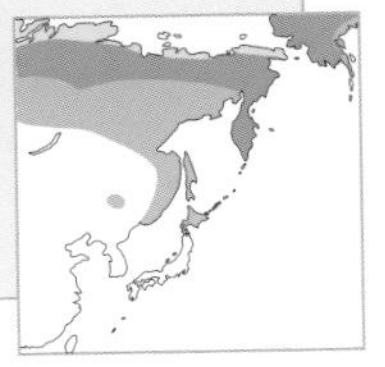

Yoshio Osawa

Adult grey morph. February, Hokkaido

Norio Yamagata

Adult grey morph. December, Hokkaido

Peregrine Falcon

Falco peregrinus ♂38cm ♀51cm WS84–120cm

Akira Yamamoto

Adult. February, Aichi

Description Large, stocky falcon with long pointed wings and characteristic black moustache on cheeks. Adult has grey upperparts, white throat, white underparts barred with grey. Female is larger than male. Immature has dark-brown upperparts, heavily streaked belly and heavily barred underwings.

Similar species Immature can be confused with Northern Hobby in flight, but Northern Hobby is smaller and slimmer, with longer narrow wings.

Range Cosmopolitan and widely distributed across temperate regions.

Status in Japan Uncommon resident throughout. Breeds on cliffs along coast.

Nobuo Shiraki

Adult. May, Nagasaki

Shuichi Tsuno

Adult. November, Wakayama

Rock Ptarmigan

Lagopus muta 37cm

Description A plump gamebird of alpine areas. Male summer plumage has dark greyish-brown head, breast and back contrasting with white underparts; red wattle over eye. Female summer plumage cryptically mottled yellowish-brown, dark brown and white. Winter plumage is entirely white except for black tail; male has black lores and red wattle over eye.
Similar species See Willow Ptarmigan.

Mitsuo Imai

Adult ♂ non-breeding. May, Toyama

Range Breeds in northern Eurasia from northern Europe east to Chukotski and Kamchatka Peninsulas, south to Transbaikalia and Kurile Islands, also in northern North America. Sedentary, forming small flocks in winter.
Status in Japan Uncommon resident in alpine dwarf stone pine areas in central Honshu. Difficult to observe in summer, but females and chicks are often seen on the Japanese Alps trails. Males are easier to find earlier in the season when they watch over their territory from conspicuous positions, such as exposed rocks.

Tokio Sugiyama

Adult ♂ breeding. June, Toyama

Akira Yamamoto

Adult ♀ breeding. August, Toyama

Willow Ptarmigan

Lagopus lagopus 38cm

Tomi Muukonen

Adult ♂ breeding. June, Varanger, Norway

Description Male has rusty-red head, neck and upper breast with red wattle over eye, white belly and wings; black tail. Female is similar to female Rock Ptarmigan, but more yellowish and irregularly spotted. Winter plumage is entirely white, apart from black tail.

Similar species Rock Ptarmigan is slightly smaller and greyer and male has black lores in winter.

Range Breeds on tundra and in taiga areas in northern Eurasia from Scandinavia east to far north-eastern Russia and Kamchatka Peninsula. In north-east Asia, breeds from lower Amur Basin and Sakhalin northwards. Sedentary and forms small flocks in winter.

Status in Japan No records.

Black Grouse

Lyrurus tetrix 40–55cm

Tomi Muukkonen

Adult ♀. August, Kuusamo, Finland

Description Male unmistakable, glossy-black with red wattle over eye. Bold white wing-bars, small white shoulder-patch and lyre-shaped tail with long curved outer-tail feathers, white undertail-coverts. Female is mottled yellowish-brown and grey overall, with pale wing-bar and slightly notched tail. In early mornings during breeding season, males get together at 'leks' and perform elaborate courtship displays.

Tomi Muukonen

Adult ♂. April, Kuusamo, Finland

Range Sedentary. Breeds in northern Eurasia from Britain east through Siberia to Amur Basin and Ussuriland, south to northern China and northern Korea. Inhabits woodlands.

Status in Japan No records.

Hazel Grouse

Tetrastes bonasia 35–37cm

Description Greyish-brown gamebird. Male has short crest, black bib with white border, greyish-brown back with dark-brown and reddish-brown barring, dark band at both sides of tail tip. Underparts are heavily barred and spotted. Female is browner with Speckled white throat.

Tadao Shimba

Adult ♂. January, Hokkaido

Range Breeds in northern Eurasia from Europe east to Okhotsk coast, south to Mongolia, north-eastern China, Sakhalin and Korea. **Status in Japan** Locally uncommon resident in coniferous and mixed forests in Hokkaido. Numbers have declined. In winter, often observed feeding in trees.

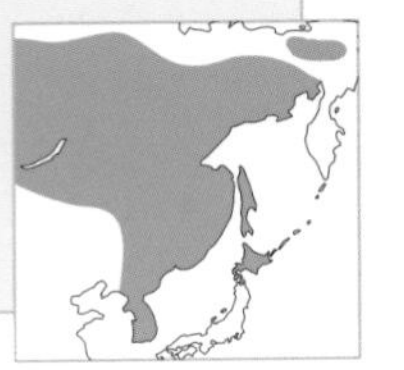

Kazuyasu Kisaichi

Adult ♀. July, Hokkaido

Himaru Iozawa

Adult ♀. June, Hokkaido

Siberian Grouse

Falcipennis falcipennis 37cm

Siegfried Klaus

Adult ♂. May, Komsomolsk, Amur, Russia

Description Male has dark grey back with dark brown barring and sparse white spotting; dark grey head with red bare skin above eyes, and black breast; black belly with heavy white spotting and white-tipped rectrices. Female dark brown above, with yellowish feather edges, and boldly white-spotted below. Usually tame and often allows close approach. Male performs displays in spring on the breeding grounds; these consist of tail-fanning, neck-stretching and erection of the red combs above the eyes.

Similar species Hazel Grouse is spotted brown and rufous below.

Siegfried Klaus

Adult ♂. May, Komsomolsk, Amur, Russia

Range Breeds in south-east Russia from eastern Siberia to the Okhotsk coast; south to Ussuriland and Sakhalin. Sedentary; inhabits dense taiga. Rare and declining.

Status in Japan No records.

Siegfried Klaus

Adult ♀. May, Komsomolsk, Amur, Russia

Black-billed Capercaillie

Tetrao urogalloides ♂95cm ♀65cm

Description Largest grouse in the region. Male unmistakable; has dark purplish-black body with heavily feathered legs; black bill, red combs above eye, and metallic green wings with white spots. Female is much smaller than male, with brown plumage heavily barred black-and-white. During courtship, males gather at leks and perform displays at dawn.
Voice Display call is a click that becomes a short trill.
Similar species Female Black Grouse is much smaller and darker, and has shorter tail.

Range Breeds from northern Siberia east to Kamchatka, south to northern Mongolia, northern China and Sakhalin. Sedentary; inhabits plains and montane coniferous forests. Population declining due to habitat destruction and hunting pressure.
Status in Japan No records.

Siegfried Klaus

Adult ♂. May, Magadan, Russia

Siegfried Klaus

Adult ♂. May, Magadan, Russia

Siegfried Klaus

Adult ♀. May, Magadan, Russia

Japanese Quail

Coturnix japonica 20cm

Teruaki Ishii

Adult ♂. March, Kyoto

Description Smallest gamebird of the region. Male has rufous-brown throat, white supercilium extending to nape and rufous-brown back streaked pale-yellow and black; orange-brown underparts with white streaks. Female is browner with dark streaks on breast and white throat.

Range Breeds in eastern Asia from northern Mongolia east to Amur Basin, Ussuriland, Sakhalin and north-eastern China, south to eastern China. Winters south to south-east Asia.
Status in Japan Uncommon winter visitor to cultivated fields, grassy fields and riverbanks. Breeds locally from central Honshu northwards. Numbers declining sharply.

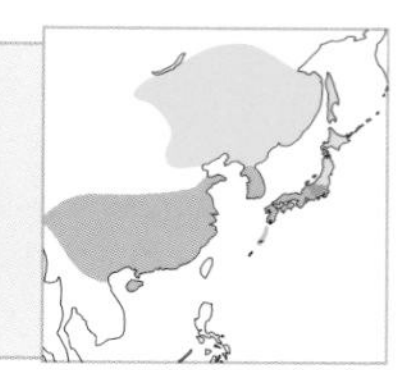

Chinese Bamboo Partridge

Bambusicola thoracica 27cm

Akira Yamamoto

Adults. October, Aichi

Description Rufous face and throat contrast with bluish-grey stripe over eye and bluish-grey breast. Grey-brown back with rufous spots; yellowish-brown underparts with dark-brown spots.
Voice Sharp and harsh *pit-pit-pit* or loud and rapidly repeated *pit-kwee.*
Similar species Female Copper Pheasant is larger and browner.

Range Resident in southern China and Taiwan.
Status in Japan Introduced from China in 1919. Common resident in woodlands and thickets in foothills from Honshu to Kyushu.

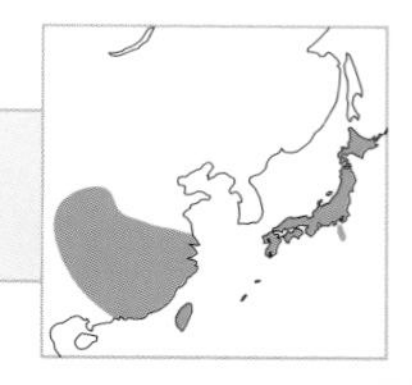

Daurian Partridge

Perdix dauurica 30cm

Description Medium-sized gamebird with conspicuous dark brown patch on belly. Orange head, brown back with dark brown barring and fine white streaks; buff-coloured under-parts with orange breast and bold dark brown barring on flanks. Female duller with less conspicuous patch on belly.
Similar species Grey Partridge *P. perdix*, widespread in Europe and western Asia, has grey underparts and brown belly patch.

Igor Karyakin

Adult ♀. June, Tuva Basin, Russia.

Range Breeds in eastern Asia from Kyrgyzstan east through Mongolia to Ussuriland and north-east China. Sedentary; inhabits open and cultivated fields. Forms small flocks outside the breeding season.
Status in Japan No records.

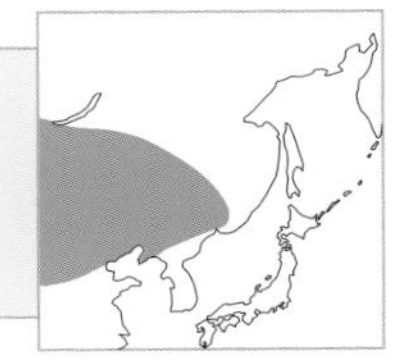

Common Pheasant

Phasianus colchicus ♂80cm ♀60cm

Description Similar to Green Pheasant, but male has distinctive white neck-ring, chestnut under-parts and dark spotted yellowish sides of breast, flanks and back. Female is difficult to separate from Green Pheasant.
Voice Male displays with loud and harsh disyllabic *korr-kock,* followed by a series of rapid wingbeats.

Tokio Sugiyama

Adult ♂. May, Tsushima

Range Breeds in central and eastern Palearctic from Caucasus east to Amur Basin and Ussuriland, south to southern China; also introduced to Europe and North America.
Status in Japan Common resident in Hokkaido and on Tsushima. Introduced to Hokkaido in 1930s. Inhabits cultivated and grassy fields.

Green Pheasant

Phasianus versicolor ♂80cm ♀60cm

Tadao Shimba

Adult ♂. April, Aichi

Tadao Shimba

Adult ♀. February, Aichi

Description Male has distinctive large patch of bare red skin on face, glossy-purple neck and glossy-green underparts. Reddish-brown shoulders are mottled black and yellow, bluish-grey wing coverts and rump; long, pointed, finely barred bluish-grey tail. Female is buff-coloured overall with heavy dark-brown markings on breast, flanks and back.

Voice Male displays with loud and harsh disyllabic *korr-kock*, followed by a series of rapid wingbeats.

Note Sometimes considered conspecific with Common Pheasant.

Range Endemic to Japan
Status in Japan Common resident from Honshu to Yakushima. Inhabits cultivated fields, riverbanks and grassy fields. The national bird of Japan.

Tadao Shimba

Adult ♂. April, Aichi

Copper Pheasant

Syrmaticus soemmerringii M125cm F55cm

Description Reddish-brown pheasant with long tail. Male has reddish-brown head and underparts, reddish-brown back mottled with black and white. Female is smaller, and is overall greyish-brown mottled with dark-brown and white. Five races in Japan, the southern race is paler; race *ijimae* in southern Kyushu has distinctive white rump. Male displays with a series of rapid wingbeats.

Akira Yamamoto

Adult ♂ *subrufus*. November, Wakayama.

Range Endemic to Japan.
Status in Japan Uncommon resident in mountain forests from Honshu to Kyushu. Feeds mainly on the ground and occasionally in trees. Easily flushed from cover.

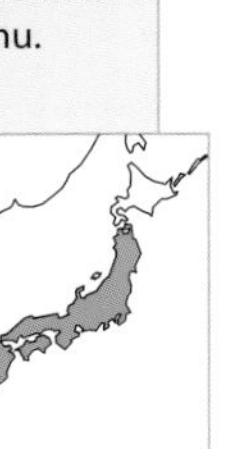

Yoshio Osawa

Immature ♂ *scintillans*. February, Yamagata

Kazuyasu Kisaichi

Adult ♀ *scintillans*. March, Yamagata

Barred Buttonquail

Turnix suscitator 14cm

Tadao Shimba

Adult ♀. July, Miyako-jima

Description Small quail-like, ground-dwelling bird. Male has white-spotted brown face, brown back mottled with black and white, black-barred yellowish-brown breast, greyish-brown belly and bluish-grey legs. Female is larger than male, with greyish head and legs and distinctive black patch from chin to upper breast. Male incubates the eggs and rears the young.

Voice Female calls in breeding season, a low booming *woo-woo.*

Nobuo Shiraki

Adult ♂. August, Okinawa-Hontou

Range Southern and eastern Asia. Breeds from India east to southern China, Taiwan and Ryukyu Islands, south to south-east Asia.

Status in Japan Common resident on southern Ryukyu Islands. Inhabits sugar cane fields and grassy fields.

Yellow-legged Buttonquail

Turnix tanki 13cm

Mike Parker

Adult ♂. May, Hebei, China.

Description Small, short-tailed bird with yellow bill and feet. Brownish upperparts with spotted wing-coverts and scapulars; underparts white with rusty patch on breast and dark spots on the flanks. Female is brighter than male with rufous patch on shoulders; male incubates eggs and rears young.

Range Breeds from Indian subcontinent east through northern south-east Asia to eastern China, north to Transbaikalia and Ussuriland. Northern population is migratory and winters in southern Asia. Rare passage migrant to Korea. Inhabits plains and grassy fields.

Status in Japan No records.

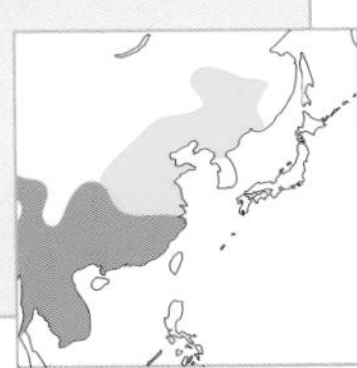

Demoiselle Crane

Anthropoides virgo 97.5cm

Description Small crane with black face, long white ear-tufts, black throat and foreneck, grey body and long, drooping tertials. Juvenile has brownish upperparts, head and neck. In flight, black primaries contrast with grey forewings and secondaries.

Range Breeds in southern Russia east to Mongolia and north-western China. Winters south to northern Africa, India and southern China.
Status in Japan Very rare passage migrant and winter visitor to rice fields and grasslands. Most often observed at Arasaki in Kagoshima.

Tokio Sugiyama

Adult breeding. May, Ishikawa

Common Crane

Grus grus 114cm

Description Dark-grey crane with black head, throat and foreneck. Red crown, broad white stripe from ear-coverts down to neck, pale yellowish bill and dark legs. Juvenile has browner head with greyish neck and body. In flight, black flight feathers contrast with grey forewings.
Similar species Hooded Crane is smaller; has white head and neck. Demoiselle Crane is smaller, has black face with long white ear-tufts.

Range Breeds in northern Eurasia from Scandinavia east through southern Siberia to Amur Basin and northern China. Winters in temperate Eurasia and northern Africa.
Status in Japan Rare winter visitor to rice fields and freshwater marshes. Small numbers winter regularly at Arasaki in Kagoshima.

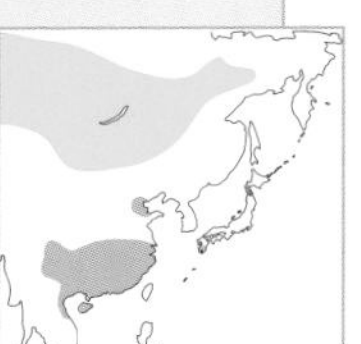

Tadao Shimba

Adult. January, Kagoshima

Siberian Crane

Grus leucogeranus 135cm

Tadao Shimba
Adult, January, Arasaki, Kagoshima pref.

Description Large white crane with bare red facial skin, long red bill and reddish legs. Juvenile has yellowish-brown head and neck, and white back tinged yellowish-brown. In flight, distinctive black primaries contrast with white wings.
Similar species Red-crowned Crane has red crown, black neck, white primaries, and black secondaries and tertials.

Tadao Shimba
Adult. January, Kagoshima

Range Breeds in northern Siberia and winters mainly in north-western India and eastern China. Endangered.
Status in Japan Very rare winter visitor to cultivated fields and freshwater marshes. Most records are from Arasaki in Kagoshima.

Sandhill Crane

Grus canadensis 100cm

Tadao Shimba
Adult. February, Nagano.

Description Greyish body tinged brown. Red forehead and dark bill and legs. Immature has rusty-brown head. In flight, black flight feathers contrast with grey forewings.
Similar species Separated from other cranes in north-east Asia by overall greyish-brown colour.

Range Breeds from far north-eastern Russia and Alaska east through Arctic Canada to Hudson Bay, south to Florida. Northern population is highly migratory and winters in southern United States.
Status in Japan Very rare winter visitor to cultivated and grassy fields. Small numbers annually winter at Arasaki in Kagoshima.

White-naped Crane

Grus vipio 127cm

Description Large greyish crane with distinctive facial and neck pattern. Red face, yellow bill, white head and neck with dark-grey stripes on side of neck, grey back with dark-grey underparts. In flight, dark flight feathers contrast with grey forewings. Juvenile has rusty mottling on head.
Similar species Hooded Crane is smaller and has red crown, white face and uniformly dark wings.

Tadao Shimba

Adult. January, Kagoshima

Range Breeds from Transbaikalia east to Amur basin, Ussuriland and north-eastern China. Winters mainly in eastern China and southern Japan.
Status in Japan Locally common winter visitor to Arasaki in Kagoshima where around 2,000 winter in most years; rare elsewhere.

Tadao Shimba

Adult and juveniles. January, Kagoshima

Tadao Shimba

Juvenile (left) and adult (right). January, Kagoshima

Hooded Crane

Grus monacha 96.5cm

Adults. January, Kagoshima

Adults and one juvenile. January, Kagoshima

Adult. January, Kagoshima

Description Adult has white head with dark forehead and red crown and dark slaty-black body, greenish bill and dark legs. Juvenile has yellowish-brown head and neck. In flight shows uniform dark wings.
Similar species Common Crane is larger with black throat and foreneck and pale wing coverts. Hybrid Common x Hooded Cranes have been observed regularly at Arasaki.

Range Breeds in southern Siberia east to Amur Basin and northern China. Winters mainly in southern Japan; uncommon passage migrant to Korea.
Status in Japan Locally abundant winter visitor to Arasaki in Kagoshima where about 10,000 birds, presumed to represent the majority of the world population, over-winter. Other regular wintering area includes Yamaguchi and Kochi; rare elsewhere.

Red-crowned Crane

Grus japonensis 140cm

Description Largest crane in Japan. White overall with red crown; black forehead, neck and long tertials, long yellow bill and black legs. In flight, black secondaries and tertials contrast with white forewings and primaries. Juvenile has yellowish-brown head and neck and white back tinged yellowish-brown.
Similar species Siberian Crane has red face and black primaries.

Range Breeds from north-east Mongolia eastwards to Amur Basin, Ussuriland and north-eastern China. Winters south to eastern China.
Status in Japan Locally common resident in eastern Hokkaido, very rare elsewhere. Winter feeding stations are a popular tourist attraction in Hokkaido. Breeds in freshwater marshes and grassy meadows.

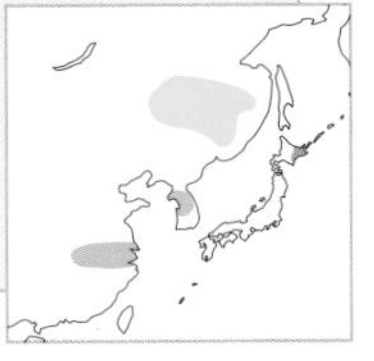

Tadao Shimba

Adult. December, Hokkaido

Tadao Shimba

Adult. December, Hokkaido

Tadao Shimba

Juvenile. December, Hokkaido

Slaty-legged Crake

Rallina eurizonoides 26cm

Kazuyasu Kisaichi

Adult. July, Miyako-jima

Description Dark olive-brown upper-parts, reddish-brown face, throat and breast. Heavily black and white barred belly and undertail-coverts, dark bill, white chin and black legs.
Voice Loud *kek-kek* and croaking *kraah* calls, often at night.
Similar species Ruddy-breasted Crake is smaller and lacks barring has shorter bill and reddish legs.

Peter and Michelle Wong

Adult. February, Hong Kong

Range Breeds from India east to southern China and south-east Asia, north to Taiwan and Ryukyu Islands.
Status in Japan Uncommon resident on Yaeyama Islands. Inhabits sub-tropical forest floor and forest edges.

Swinhoe's Rail

Coturnicops exquisitus 13cm

Shigeo Ozawa

Adult. January, Chiba

Description Smallest rallid in the region. Brown back with dark stripes and fine white bars; white spotting on breast and white belly. White secondaries are distinctive.

Range Breeds in Transbaikalia, north-east China and Ussuriland. Winters south to southern China. Secretive.
Status in Japan Rare winter visitor to freshwater marshes and rice fields. Breeding has recently been confirmed in Aomori.

Water Rail

Rallus aquaticus 29cm

Description Olive-brown back with dark brown streaks. Bluish-grey face with dark eye-stripe, long slightly decurved reddish bill with brown upper mandible, greyish breast with black and white barred flanks.
In breeding plumage, bill becomes bright red.
Similar species Baillon's Crake is much smaller and short-billed. Ruddy-breasted Crake has dark greenish-brown back and reddish-brown underparts.

Akira Yamamoto

Adult breeding. March, Aichi

Yoshihito Goto

Adult non-breeding. February, Aichi

Akira Yamamoto

Adult non-breeding. January, Osaka

Range Breeds across temperate Eurasia from Europe east through southern Siberia to north-east China, Ussuriland and Sakhalin, south to Korea and northern Japan. Winters south to northern Africa, southern and south-east Asia.
Status in Japan Breeds in northern Japan and common winter visitor to warmer regions. In winter, inhabits rice fields, reed beds and fresh-water marshes.

White-breasted Waterhen

Amaurornis phoenicurus 32.5cm

Yoshihito Goto

Adult. August, Ishigaki-jima

Description Unmistakable black–and–white long-legged waterbird. Sooty-black upperparts, becoming dark-brown from lower back to tail. White face and underparts with chestnut undertail-coverts, yellowish-green bill with red base to upper mandible, and yellowish-green legs. Juvenile duller.

Voice Loud, monotonous and repeated *kruk-kruk-kruk,* followed by a variety of roars and grunts.

Tokio Sugiyama

Immature. March, Okinawa-Hontou

Range South and eastern Asia. Breeds from Indian subcontinent east to China, Taiwan and southern Japan, south to south-eastern Asia. Northern population is migratory. Rare visitor to southern Korea.

Status in Japan Common resident on Ryukyu Islands, rare elsewhere. Range seems to be expanding north. Inhabits rice fields, mangrove swamps, riverbanks and freshwater marshes.

Tokio Sugiyama

Immature. March, Okinawa-Hontou

Okinawa Rail

Gallirallus okinawae 30cm

Description Flightless. Rich olive-brown upperparts, large bright red bill, black face with white line from rear of eye across ear-coverts to side of neck, black underparts finely barred white, red legs.
Voice Loud and rapidly repeated *kek-kek-kek-kek* calls at night.

Mitsuo Imai

Adult. January, Okinawa-Hontou

Range Endemic to Okinawa-Hontou in Japan.
Status in Japan Discovered in 1981. Restricted to subtropical forests in northern Okinawa-Hontou. Skulks on the forest floor and is often difficult to detect. Roosts in trees at night. Threatened.

Mitsuo Imai

Adult. December, Okinawa-Hontou

Akira Yamamoto

Adult. February, Okinawa-Hontou

Baillon's Crake

Porzana pusilla 12.5cm

Tokio Sugiyama

Adult. September, Aichi

Tokio Sugiyma

Adult. September, Aichi

Description Small crake with bluish-grey face, breast and upper belly. Olive-brown back with black and white streaks, black and white barred lower belly and undertail-coverts, brown eye-patch, short greenish-yellow bill and greenish legs.
Similar species Swinhoe's Crake is much smaller and has white streaks on face and white secondaries.

Range Widely distributed across temperate Eurasia, south to southern Africa and Australia. Northern population is migratory. In north-east Asia, breeds in Ussuriland, Amur Basin, north-eastern China and northern Japan; migrates in autumn to warmer regions.
Status in Japan Rare summer visitor to northern Honshu and Hokkaido. Breeds in shallow freshwater marshes. Rare passage migrant elsewhere with a few winter records in southern Japan.

Sumit Sen

Adult. January, Kolkata, india

Ruddy-breasted Crake

Porzana fusca 22.5cm

Description Medium-sized crake with dark, olive-brown back and reddish-brown face, breast and upper belly. Short, dark, slightly decurved bill, pale chin, black–and–white barred lower belly and undertail-coverts, red legs. Becomes duller in non-breeding plumage. Two races in Japan; *phaeopyga* on Ryukyu Islands is slightly darker than *erythrothorax* on other islands.
Voice Short, repeated and accelerating *kyot-kyot-kyot.* Calls at night during breeding season.
Similar species Slaty-legged Crake is larger, has darker legs and black–and–white barred belly. Band-bellied Crake has variable amounts of white on wing-coverts and more boldly barred belly.

Range Southern and eastern Asia. Breeds from Pakistan east to eastern China and Taiwan, south to Indonesia, north to north-east China and Japan. Northern population is migratory. Uncommon summer visitor to Korea.
Status in Japan Uncommon summer visitor to rice fields, riverbanks and freshwater marshes. Some winter in southern Japan. Race *phaeopyga* is resident on Ryukyu Islands.

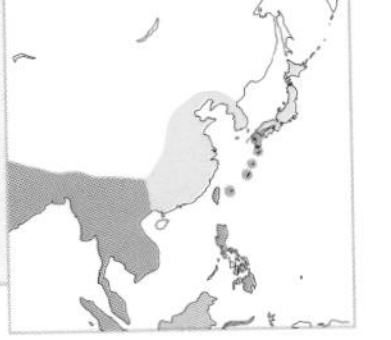

Tokio Sugiyama

Adult breeding. June, Aichi

Akira Yamamoto

Adult non-breeding. January, Osaka

Akira Yamamoto

Adult breeding. June, Aichi

Band-bellied Crake

Porzana paykullii 22cm

Jung-jang Jo

Adult. August, Sosan, South Korea

Description Olive-brown, upperparts with variable white-marked greater and median-coverts. Rufous face and breast, white throat and boldly black–and–white barred belly and undertail-coverts. Red legs.
Similar species See Ruddy-breasted Crake.

Range Breeds in north-east Asia from southern Ussuriland south to north-east China. Winters in south-east Asia. Inhabits drier places than other crakes.
Status in Japan Vagrant. First recorded in May 1993 in Hokkaido. Few other sightings.

Watercock

Gallicrex cinerea 33cm

Hyun-tae Kim

Adult ♂ breeding. June, Sosan, South Korea

Description Male breeding is sooty-black overall with conspicuous red frontal shield, buff-coloured edges to back feathers and buff-coloured undertail-coverts, yellow bill and reddish legs. Female and non-breeding male have dark-brown crown and upperparts with buff-coloured edges to mantle, buff-coloured underparts, yellow bill and dull greenish legs.
Voice Loud and repeated *kok-kok-kok* or *kruk-kruk-kruk*.
Similar species Common Moorhen is slightly smaller, has uniform dark brown back, white flank line and yellow legs.

Range Southern and eastern Asia. Breeds from Pakistan east to eastern China and Taiwan, north to north-east China and southern Ussuriland. Uncommon summer visitor to Korea.
Status in Japan Uncommon resident on Yaeyama Islands. Rare passage migrant elsewhere. Freshwater marshes, rice fields and reedbeds.

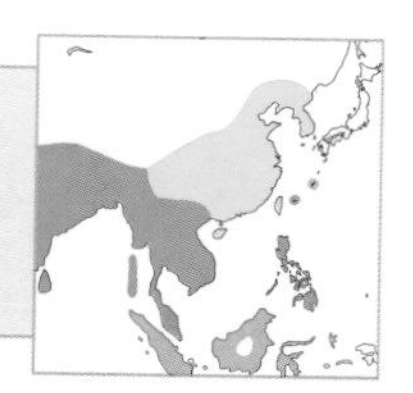

Common Moorhen

Gallinula chloropus 32.5cm

Description Plump dark waterbird, with sooty-black face, breast and underparts. Dark olive-brown back, red frontal shield and yellow-tipped red bill, white flank stripes, white undertail-coverts and yellow legs. Juvenile duller.
Voice Loud and explosive *krrrruk* and wide variety of clucking and chattering calls.
Similar species Eurasian Coot has white bill and frontal shield, and lacks white flank stripe.

Adult. January, Osaka

Range Widespread throughout temperate and tropical Eurasia, also in Africa and both North and South America. In north-east Asia, an uncommon summer visitor from Ussuriland southwards.
Status in Japan Common resident in reedbeds, ponds and rice fields. Northern populations move to warmer areas in winter.

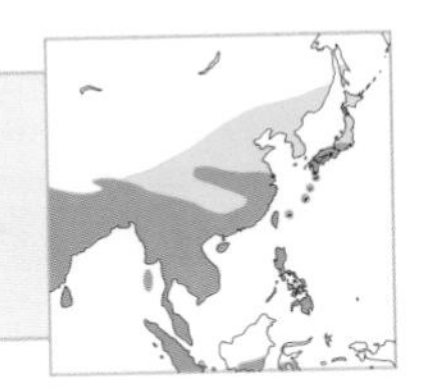

Eurasian Coot

Fulica atra 39cm

Description Uniform sooty-black waterbird, with striking white bill and white frontal shield; silvery-grey legs. Immature has greyish-brown back and whitish underparts, greyish bill. In flight, white trailing edges to secondaries.
Voice Noisy, harsh *kik*.
Similar species Common Moorhen has red bill and frontal shield, white flank stripe and white undertail-coverts.

Adult. June, Aichi

Range Widespread in Eurasia, breeds from Europe east to Okhotsk coast and Sakhalin, south to Indian subcontinent, northern Mongolia and northern China; also in north Africa and Australia. Winters south to tropical regions.
Status in Japan Uncommon resident. Breeds mainly in northern Japan and winters from Honshu southwards. Inhabits freshwater marshes and ponds. Often forms large flocks in winter.

Great Bustard

Otis tarda 100cm

Tokio Sugiyama

Adult ♂. March, Aichi

Description Large, heavily-built bird with thick neck. Grey head and neck, yellowish-brown back with fine black barring, white wing coverts, whitish underparts and pale-brown legs. Male is larger than female and in breeding plumage has long white moustaches and chestnut breast-band.

Range Breeds sparsely in Eurasia from Iberian Peninsula east through central Asia to Mongolia and north-west China. Winters south to temperate regions. Inhabits steppes, deserts and grassy fields. Sharply declining. In north-east Asia, bred in western part of Heilongjiang and Jilin until 1990s but now only small flocks occur in winter.
Status in Japan Vagrant. Occurs in cultivated and grassy fields.

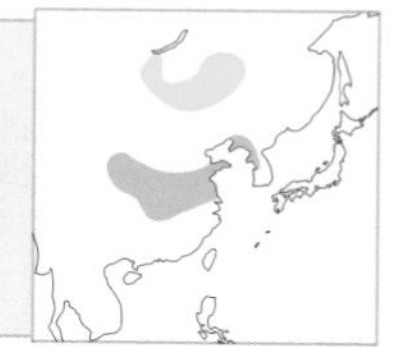

Pheasant-tailed Jacana

Hydrophasianus chirurgus 55cm

Tokio Sugiyama

Adult breeding. July, Kagawa

Description Unmistakable. Extremely long toes and claws adapted for walking on floating vegetation. White head and neck with black-bordered yellow patch on nape, dark-brown back, white wings with dark wing tips, dark chocolate-brown underparts and long black tail. Juvenile and non-breeding plumages have brown crown and hindneck, dark line from eye down side of neck onto breast, brown back and white underparts; also lack long tail feathers.

Akira Yamamoto

Juvenile. October, Miyako-jima

Range Southern and eastern Asia. Breeds from Indian subcontinent east to southern China and Taiwan, south to Indonesia.
Status in Japan Rare, mainly summer-autumn visitor to southern Japan. Mostly seen in lotus ponds and fresh-water marshes.

Greater Painted-snipe

Rostratula benghalensis 23.5cm

Description Medium-sized, snipe-like wader with distinctive facial pattern. Female is larger and more brightly coloured than male. Breeding plumage female has rich chestnut face, neck and upper-breast. White line separates wing from breast, neck and mantle. Also diagnostic white patch behind eye. Breeding plumage male is paler than female overall with yellowish upperwing-coverts and yellow patch behind eye. Non-breeding plumage female lacks chestnut colour and has uniform yellowish-brown back.
Voice Female calls at night in breeding season. Soft, hooting *koh-koh-koh*.

Akira Yamamoto

Adult ♀ breeding. June, Aichi

Range Resident in Africa, Australia and southern Asia; east to eastern China and southern Japan.
Status in Japan Uncommon resident in rice fields and shallow freshwater marshes from central Honshu southwards. Northern populations move south in winter.

Akira Yamamoto

Adult ♀ breeding. June, Aichi

Akira Yamamoto

Adult ♂ breeding. August, Aichi

Eurasian Oystercatcher

Haematopus ostralegus 45cm

Tokio Sugiyama

Adult. November, Aichi

Description Unmistakable, large black and white wader with distinctive orange-red bill and pink legs. Immature has brownish-black back. **Voice** Sharp and repeated *pit-pit* or sharp whistled *kleep*. The race in eastern Asia, *osculans*, is considered a separate species (Eastern Oystercatcher *H. osculans*) by some authorities.

Akira Yamamoto

Adult. October, Ishikawa

Range Breeds across Eurasia east to Kamchatka Peninsula and north-eastern China. Northern populations are migratory and winter south to northern Africa, Middle East and southern Asia. **Status in Japan** Uncommon winter visitor to coastal mud-flats and sandy shores. Numbers have increased in recent years.

Akira Yamamoto

Adult. October, Ishikawa

Black-winged Stilt

Himantopus himantopus 31cm

Description Unmistakable, long-legged wader with black wings and back, white underparts; straight, thin black bill and very long pinkish-red legs. Female browner above. Some males have black crown and nape. Juvenile has brown crown, neck and back.
Voice Sharp *kek-kek* particularly when disturbed.

Akira Yamamoto

Adult. November Aichi

Range Widely distributed in central Eurasia, Africa, southern Asia and Australia.
Status in Japan Uncommon passage migrant and winter visitor, mainly to central-Honshu southwards. Some breed in freshwater marshes in central Honshu. Common winter visitor to Okinawa Islands.

Akira Yamamoto

Juvenile. November, Aichi

Akira Yamamoto

Adult (left) and juvenile (right). November, Aichi

Oriental Pratincole

Glareola maldivarum 26.5cm

Tadao Shimba

Adult breeding. May, Tsushima

Description Distinctive tern-like wader with long pointed wings, short forked tail and short black legs. Olive-brown overall with creamy-coloured throat, bordered with black line, in breeding plumage. In non-breeding plumage, throat becomes duller bordered with indistinct black streaks. In flight shows chestnut underwing-coverts and white rump.

Tadao Shimba

Adult non-breeding. September, Aichi

Range Breeds from India east to eastern China. Northern populations are migratory and winter south to northern Australia.

Status in Japan Uncommon passage migrant on cultivated fields, rice fields and rough ground. Breeds locally in southern Japan.

Akira Yamamoto

Adult breeding. May, Miyako-jima

Pied Avocet

Recurvirostra avosetta 43cm

Description Elegant black and white wader with long, tapering, upcurved bill and bluish-grey legs. Black cap, hindneck and white underparts. Diagnostic bold black and white pattern on upperwings. Sexes are alike, immature browner.
Voice Soft *kluit.*

Mitsuo Imai

Adult ♂. March. Ibaragi

Range Breeds in Eurasia east to Transbaikalia area. Inland population migrates to southern Europe, Africa and the Middle East, southern China, Taiwan and Korea.
Status in Japan Rare passage migrant and winter visitor to mainly Honshu and southward. More frequently seen in Ryukyu Islands.

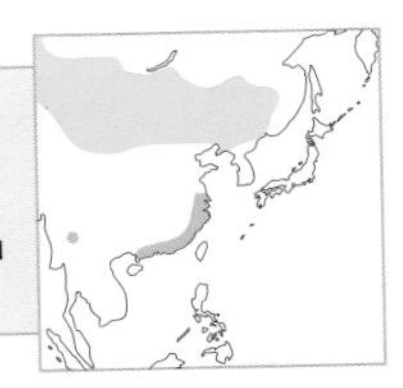

Northern Lapwing

Vanellus vanellus 31.5cm

Description Large plover with unmistakable facial pattern, long, thin crest, dark glossy-green upperparts and white belly. Has broad, rounded wings. In breeding plumage, black breast extends onto throat. Immature has buff-coloured fringes on upperparts. Slow and deep wingbeats.
Voice Distinctive whistled *myu-wit.*

Tadao Shimba

Adult non-breeding. March, Aichi

Range Breeds across temperate Eurasia east to south-eastern Russia and north-eastern China. Northern and inland populations are migratory. Common winter visitor to southern Korea.
Status in Japan Common winter visitor from Honshu southwards. Prefers rice fields and riverbanks. There are a few breeding records.

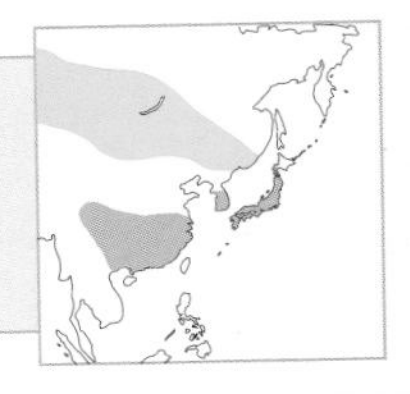

Grey-headed Lapwing

Vanellus cinereus 35.5cm

Tadao Shimba

Adult. January, Aichi

Description Large, noisy plover with long yellow legs. Bluish-grey head and breast with dark breast-band. Greyish-brown back, white tail with dark subterminal band, yellow bill with black tip. In flight, dark primaries contrast with white secondaries.

Voice Loud, noisy and explosive *kik-kik* when disturbed in breeding season.

Tadao Shimba

Adult. October, Mie

Range Breeds in north-east China and Japan. Winters in southern China and northern parts of south-east Asia. Rare passage migrant to Korea.

Status in Japan Common resident from central to northern Honshu. Rare elsewhere. Breeds in rice fields and cultivated fields.

Tadao Shimba

Adult. May, Aichi

Pacific Golden Plover

Pluvialis fulva 24cm

Description Medium-sized plover with long dark legs and dark bill. In breeding plumage, golden-yellow spangled back and black underparts separated by white band across flanks. In non-breeding plumage has buffish supercilium, grey-brown upperparts and buffish-brown underparts, streaked darker.
Voice Rapid whistling *tloo-ee*.
Similar species Non-breeding Grey Plover is larger, greyer above and has heavier bill; in flight shows white rump and distinctive black axillaries. See also American Golden Plover.

Range Breeds across arctic Siberia and in western Alaska. Winters in southern Asia, Australia and tropical Pacific. Uncommon passage migrant in north-east Asia.
Status in Japan Common passage migrant in rice fields and coastal mud-flats. Some winter in southern Japan.

Tadao Shimba

Adult ♂ breeding. May, Aichi

Akira Yamamoto

Adult non-breeding. December, Aichi

Akira Yamamoto

Adult non-breeding. December, Aichi

American Golden Plover

Pluvialis dominica 23–28cm

Kazuyasu Kisaichi

Adult ♂ breeding. June, Churchill, Canada

Tadao Shimba

Adult breeding (in moult). August, Ontario, Canada

Description Very similar to Pacific Golden Plover, but has slightly shorter pointed bill, slightly shorter legs, longer wings and shorter tertials. At rest, four outer primaries extend beyond tail. Some Pacific Golden Plovers have long wings, but normally only two outer primaries extend beyond tail.
Voice High-pitched, shrill *chu-wheep,* thinner than call of Pacific Golden Plover.
Similar species Non-breeding Grey Plover is larger greyer above and has heavier bill; in flight shows white rump and black axillaries. See also Pacific Golden Plover.

Range Breeds across arctic North America and winters in South America.
Status in Japan Has been reported several times, but there is only one reliable record, from Saitama in April 1987.

Tadao Shimba

Adult breeding (in moult). August, Ontario, Canada

Grey Plover

Pluvialis squatarola 29.5cm

Other name Black-bellied Plover
Description Stout plover with dark legs and black bill. Unmistakeable in breeding plumage; black belly contrasts with spangled black and white back. In non-breeding plumage has grey back and white belly. In flight shows white uppertail-coverts and distinctive black axillaries.
Voice Distinctive, loud and far-carrying slurred whistle *tlee-oo-ee.*
Similar species Pacific Golden Plover is smaller, has grey axillaries and lacks contrasting rump.

Norio Yamagata

Adult ♂ breeding. May, Aichi

Range Breeds across arctic Eurasia and North America and moves south in winter. Common winter visitor to southern Korea.
Status in Japan Common passage migrant to coastal mudflats and many winter from Honshu southwards. Rarely seen at inland waters.

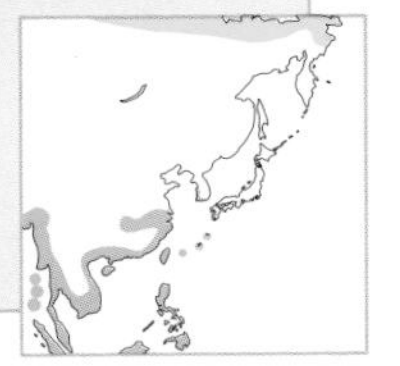

Tadao Shimba

Adult ♀ breeding. May, Aichi

Tadao Shimba

Juvenile. September, Mie

Common Ringed Plover

Charadrius hiaticula 19cm

Akira Yamamoto

Adult breeding. August, Aichi

Description Small plover with greyish-brown back and white underparts. Similar to Little Ringed Plover but slightly larger, has broader breast-band, faint yellow eye-ring and clear white wing-bar in flight. Orange at base of bill in breeding plumage. Legs orange.
Voice Low-pitched, rising *pooh-eep,* distinct from that of Little Ringed Plover.
Similar species See Long-billed and Little Ringed Plovers.

Tokio Sugiyama

Adult non-breeding. December, Aichi

Range Breeds across northern Eurasia east to Chukotski Peninsula. Winters mainly in Africa, Europe and western Asia.
Status in Japan Rare passage migrant to coastal mudflats but small numbers regularly winter on Shiokawa tidal-flats in Aichi.

Akira Yamamoto

Adult non-breeding. December, Aichi

Long-billed Plover

Charadrius placidus 21cm

Description Small-medium plover with greyish-brown back and white underparts. Similar to Little Ringed Plover, but has longer bill, longer legs and greyish-brown eye-stripe. Shows faint yellow eye-ring and indistinct white wing-bar in flight.
Voice Clear whistled *pewee.*

Range Breeds from Ussuriland south to north-east China and Korea. Winters in southern China and south-east Asia.
Status in Japan Uncommon resident from Kyushu northwards. Northern population moves south in winter; rare winter visitor on Ryukyu Islands. Prefers inland freshwater habitats and rarely occurs in coastal areas. Breeds on stony riverbanks.

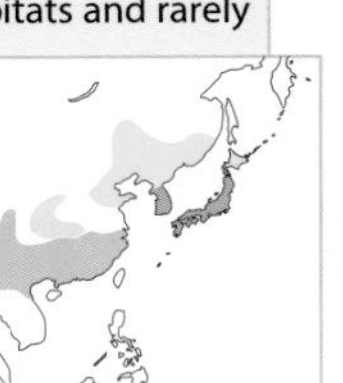

Akira Yamamoto

Adult non-breeding. February. Aichi

Akira Yamamoto

Adult non-breeding. February. Aichi

Nobuo Shiraki

Adult breeding. May, Osaka

Little Ringed Plover

Charadrius dubius 16cm

Akira Yamamoto

Adult ♂ breeding. May, Aichi

Description Small plover with greyish-brown back and white underparts. Dark bill, yellow eye-ring, pinkish legs and black breast-band. No wing-bar on upperwing. In non-breeding and juvenile plumages, has duller markings and faint eye-ring.
Voice Repeated *creeh-creeh* often heard during display flight in breeding season; also musical *pee-oh* call.
Similar species See Common Ringed Plover and Long-billed Plover.

Tadao Shimba

Juvenile. October, Aichi

Range Breeds across temperate and tropical Eurasia and winters in Africa, Indian subcontinent and south-east Asia. Common summer visitor to north-east Asia.
Status in Japan Common summer migrant. Breeds in cultivated fields, riverbanks and stony open fields. A few winter in southern Japan.

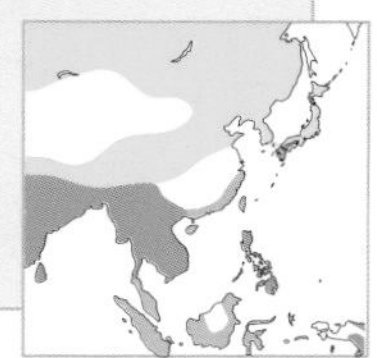

Tadao Shimba

Adult ♂ breeding. April, Aichi

Kentish Plover

Charadrius alexandrinus 17cm

Description Small plover with broken breast-band, black bill and dark legs. In breeding plumage, male has rufous crown with black mark on forehead. In non-breeding plumage has duller head markings and breast-band. In flight shows white wing-bar on upperwings.
Voice Soft *twit.*
Similar species Lesser Sand Plover is larger, has a stouter bill and lacks white collar.

Tadao Shimba

Adult ♂ breeding. March, Mie

Range Breeds across temperate northern hemisphere. Northern populations are migratory. Common summer visitor to north-east Asia north to Ussuriland.
Status in Japan Common resident. Breeds in sandy coastal areas. Northern population moves south in winter. Declining due to loss of breeding habitat.

Tadao Shimba

Adult ♀ breeding. March, Mie

Tadao Shimba

Adult non-breeding. April, Ishigaki-jima

Lesser Sand Plover

Charadrius mongolus 19.5cm

Mitsuo Imai

Adult ♂ breeding. April, Chiba

Akira Yamamoto

Adult non-breeding. October, Aichi

Description Small to medium-sized plover, with greyish-brown upperparts, white underparts, dark bill and dark grey legs. In breeding plumage has broad chestnut breast-band to hindneck and black eye-stripe. Female is duller. In non-breeding plumage has grey back, white underparts and broken breast-band. In flight shows white wing-bar.
Voice Soft trilling *chitik*.
Similar species Greater Sand Plover is larger, has longer bill and longer legs. Leg colour varies in both species and is not always reliable as a field mark.

Range Breeds in central Asia, eastern Siberia, and Kamchatka and Chukotski Peninsulas. Winters in East Africa, southern Asia and Australia. Common migrant in coastal areas in north-east Asia.
Status in Japan Common passage migrant to mudflats and sandy coasts. Some winter in southern Japan.

Teruaki Ishii

Juvenile. September, Shizuoka

Greater Sand Plover

Charadrius leschenaultii 21.5cm

Description Medium-sized plover, with greyish-brown upperparts and white underparts. Long dark bill and long yellowish legs. Leg colour mostly yellowish, but some have dark legs. In breeding plumage has rufous breast-band to hindneck and black eye-stripe; rufous breast-band narrower than in Lesser Sand Plover. In non-breeding plumage has grey back, white underparts with broken breast-band. In flight shows white wing-bar.
Voice Short melodious trill *trrri.*
Similar species See Lesser Sand Plover.

Range Breeds from Turkey east through central Asia to Russian Altai. Winters in East Africa, southern Asia and Australia.
Status in Japan Uncommon passage migrant on mudflats and sandy coasts. Common on Ryukyu Islands where some winter.

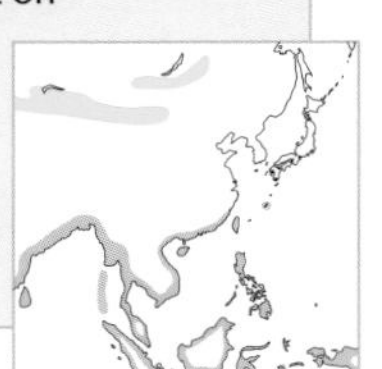

Tadao Shimba

Adult breeding. April, Ishigaki-jima

Tokio Sugiyama

Juvenile. September, Aichi

Akira Yamamoto

Adult non-breeding. September, Miyako-jima

Oriental Plover

Charadrius asiaticus 22.5cm

Tokio Sugiyama

Adult breeding. April, Kyoto

Description Medium-sized plover, with greyish-brown upperparts and white underparts, black bill and long yellow legs. In breeding plumage, white face contrasts with rufous breast. Male has black margin on the lower edge of rufous breast. In non-breeding plumage brownish above with broad, buffish eye-stripe. Wing-bar indistinct.
Voice Repeated sharp whistle *chip-chip-chip*.
Similar species Pacific Golden Plover has golden-brown upperparts. Greater Sand Plover has different facial pattern, a shorter neck and a prominent wing-bar.

Akira Yamamoto

Juvenile. August, Miyakojima

Range Breeds in southern-central Siberia, Mongolia and north-eastern China. Winters in south-east Asia and northern Australia.
Status in Japan Rare passage migrant to meadows and fields. Mostly recorded in south-western Japan in spring and seen annually on Yonaguni-jima.

Tadao Shimba

Adult non-breeding. May, Tsushima

Eurasian Dotterel

Eudromias morinellus 21cm

Description Medium-sized plover, greyish-brown above with buffish feather edging and white underparts; black bill and yellow legs, broad pale supercilium meeting on nape, and narrow white breast-band. In breeding plumage has chestnut belly and greyish-brown breast. Juveniles have bold pale feather edging above.
Similar species Pacific Golden Plover is larger, lacks breast band has dark legs, golden yellow upperparts and supercilium does not reach the nape.

Shigeo Ozawa

Juvenile. September, Chiba

Range Breeds across arctic Eurasia and on mountains of inland Siberia. Winters in Middle East and North Africa.
Status in Japan Very rare passage migrant on sandy coasts, cultivated fields and highland meadows. Mostly recorded in autumn.

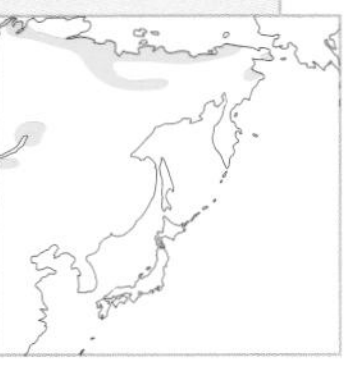

Tadao Shimba

First-winter. November, Mie

Tadao Shimba

First-winter. November, Mie

Eurasian Woodcock

Scolopax rusticola 34cm

Tokio Sugiyama

Adult. February. Aichi

Description Medium-large and long billed, with brown barred plumage, dark lores, white eye-ring and dark bars on head. Heavily-built body and short legs gives stocky appearance. Broad, rounded wings with upperwing-coverts tipped whitish.
Voice High thin *chi-kwik* mostly heard during display flights at dawn and dusk.
Similar species Amami Woodcock on Amami and Okinawa Islands.

Akira Yamamoto

Adult. December. Aichi

Range Breeds across Eurasia east to Ussuriland, Sakhalin and north-east China. Northern population is migratory. Uncommon visitor to Korea.
Status in Japan Breeds in deciduous forests in northern Honshu and Hokkaido. In winter, visits rice fields and grassy fields from Honshu southward.
Seldom seen during daytime; often seen at night feeding in open fields.

Toshihiro Takada

Adult. April, Fukui

Amami Woodcock

Scolopax mira 36cm

Description Resembles Eurasian Woodcock, but has slightly shorter bill and strong, slightly longer legs. Some individuals have diagnostic pink skin exposed around eye and indistinct eye-ring. Dark rufous brown above.
Similar species Eurasian Woodcock has a longer tail with a dark subterminal band and has less uniform upperparts. It occurs alongside Amami Woodcock in winter.

Shigeo Ozawa

Adult. September, Amami-Oshima

Range Endemic to Amami and Okinawa Islands in Japan.
Status in Japan Locally common on Amami Island and often found at night by spot-lighting along unsealed roads in forested areas. Rare on Okinawa Island.

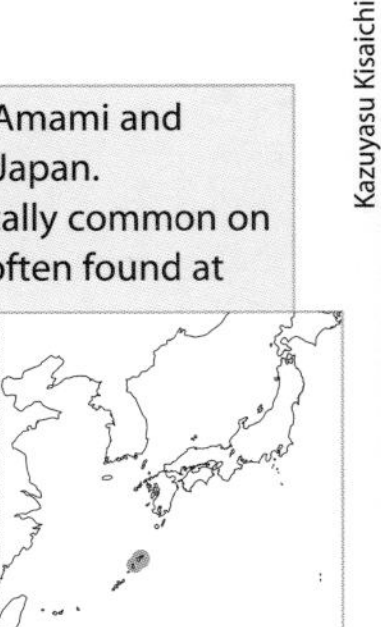

Kazuyasu Kisaichi

Adult. February, Amami-Oshima

Yoshihito Goto

Adult. February, Amami-Oshima

Solitary Snipe

Gallinago solitaria 30cm

Shigeo Ozawa

Adult. February, Kanagawa

Shigeo Ozawa

Adult. February, Kanagawa

Description Large, dark-coloured snipe with dark-brown breast and heavily barred underparts. White edges of back and shoulder feathers form clear white lines. In flight, shows indistinct white-trailing edge to wings.
Voice Harsh, loud *jhehet.*
Similar species Solitary Snipe is darker than other snipes and has much more heavily barred underparts.

Range Breeds in mountains in the Himalayas and in central Siberia, south-eastern Russia and north-eastern China. Partially migratory, moves to lower altitudes in winter.
Status in Japan Uncommon winter visitor to mountain streams, where it is difficult to detect. Normally solitary. Rare in southern Japan.

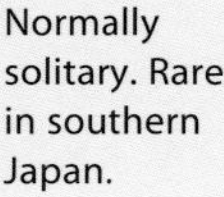

Akira Yamamoto

Adult. February, Kyoto

Latham's Snipe

Gallinago hardwickii 30cm

Description Larger than other similar snipes and proportionally more robust. Generally similar to Swinhoe's Snipe, but longer-tailed and has paler wing–coverts and tertials. Number of rectrices range from 16-18; outer tail feathers normally bright orange. In flight shows dark underwings. Normally flies low for a short distance when flushed.
Voice Harsh low *gwak*. Performs diagnostic display flight in breeding grounds giving loud, harsh *zubya-ku* followed by loud drumming sounds.
Similar species Swinhoe's Snipe is smaller, darker overall and has darker outer-tail feathers. Also see Common Snipe and Pintail Snipe.

Range Breeds in Japan, Kurile Islands and Sakhalin and winters in Australia.
Status in Japan Common summer visitor to freshwater marshes and grassy fields from central Honshu northwards. Arrives in April and departs by early September when other snipes arrive.

Teruaki Ishii

Adult breeding. May, Nagano

Tsunehiro Komai

Adult breeding. September, Aichi

Tadao Shimba

Adult breeding. July, Hokkaido

Pintail Snipe

Gallinago stenura 25cm

Tokio Sugiyama

Juvenile. September, Aichi

Description Medium-sized snipe. Slightly shorter bill, short tail and proportionally large eye gives a robust appearance compared to other snipes. Mottled black and brown back feathers delicately edged buff in adults and white in juveniles. Number of rectrices averages 26; outer six to eight feathers are narrow and needle-like but impossible to see in the field. In flight, shows faint white trailing-edges on secondaries, and dark underwing.
Voice Nasal *jhet* almost identical to Common Snipe.
Similar species See other snipes.

Tadao Shimba

Juvenile. September, Okinawa-Hontou

Range Breeds in eastern Siberia and winters from India east to southern China; south to northern Australia. Uncommon passage migrant to north-eastern Asia.
Status in Japan Rare; mainly autumn passage migrant to rice fields and freshwater marshes. Regularly visits the Ryukyu Islands.

Tadao Shimba

Juvenile. September, Okinawa-Hontou

Swinhoe's Snipe

Gallinago megala 27cm

Description Medium-sized snipe. Difficult to separate from Latham's, but slightly smaller, darker overall with dark mantle and wing-coverts. Number of rectrices averages 20, but ranges from 18–26. Outer tail feathers are dark-brown, but difficult to see in the field. In flight, lacks pale trailing-edge to secondaries. Has dark underwings.
Voice Nasal harsh *shrek*.
Similar species See other snipes.

Tokio Sugiyama

Juvenile. September, Aichi

Range Breeds from south-central Siberia east to Ussuriland and winters from India east to southern China; south to northern Australia. Uncommon passage migrant to north-east China and Korea.
Status in Japan Uncommon passage migrant in rice fields and fresh-water marshes.

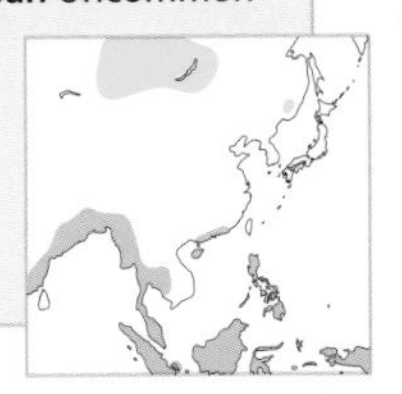

Tsunehiro Komai

Juvenile. September, Aichi

Tsunehiro Komai

Juvenile. September, Aichi

Common Snipe

Gallinago gallinago 27cm

Tadao Shimba
Adult. April, Aichi

Akira Yamamoto
Adult. April, Aichi

Description Medium-sized wader with long straight bill, distinctive striped head pattern and white belly with dark bars. Dark-brown back with creamy edges of back feathers forming clear lines on back. Brown neck and breast with fine dark streaks. In flight, shows bold white trailing-edges to secondaries and white underwing-coverts.
Voice Nasal *jhet*.
Similar species Snipe are a challenge to identify and always difficult to separate. Common Snipe can be separated from other snipes by proportionally longer bill, broader lores, narrower supercilium and broad pale straight lines on back and flight pattern.

Range Breeds across northern Eurasia east to north-eastern Russia and north-eastern China. Moves south in winter. Common passage migrant in Korea.
Status in Japan The most common snipe. Passage migrant to rice fields and freshwater marshes. Abundant in autumn. Many winter from Honshu southwards.

Tadao Shimba
Adult. January, Aichi

Long-billed Dowitcher

Limnodromus scolopaceus 30cm

Description Medium-sized, stocky wader with long straight bill and short greenish legs. Snipe-like in general appearance. In breeding plumage has reddish-brown face and breast, and belly is irregularly spotted and barred black; white eye-ring and pale supercilium, mottled black and rufous back with white feather tips. In non-breeding plumage has greyish-brown back, head and breast. White belly with dark barring. In flight shows white back and trailing–edge to wing; rump, uppertail-coverts and tail whitish barred black. Juvenile has rufous-fringed mantle and scapulars and buff-coloured underparts.
Voice Sharp *kerk-kerk.*
Similar species See Asian Dowitcher.

Tadao Shimba

Adult breeding. April, Aichi

Range Breeds in north-eastern Russia and in Alaska. Winters in coastal North and Central America.
Status in Japan Rare visitor to coastal mud-flats, rice fields and freshwater marshes. Some regularly winter at Aichi and Ibaragi.

Tokio Sugiyama

Adult breeding. August, Aichi

Tokio Sugiyama

Adult non-breeding. November, Aichi

Asian Dowitcher

Limnodromus semipalmatus 33cm

Mitsuo Imai

Adult breeding. May, Yonaguni-jima

Description Medium-large wader with long black bill and black legs. In breeding plumage, male has rufous head and breast, white eye-ring; mantle, scapular and tertial feathers blackish-brown with chestnut and white fringes. Female duller. Juvenile has white supercilium and dark brown crown; upperparts dark brown with buff fringes. Non-breeders are greyish with brown-barred flanks.
Voice Trilled whistle *kerk-kerk.*
Similar species Black-tailed Godwit is larger and has distinctive wing pattern. Long-billed Dowitcher is smaller, with shorter greenish legs.

Tokio Sugiyama

Juvenile. November, Aichi

Range Breeds in western Siberia, southern shore of Lake Baikal and north-eastern China. Winters in south-east Asia and northern Australia.
Status in Japan Rare passage migrant to coastal mudflats.

Norio Yamagata

Juvenile. August, Aichi

Black-tailed Godwit

Limosa limosa 38.5cm

Description Large wader with long black legs and long, straight bill with pale basal half and dark tip. Female is larger. In breeding plumage, male has rufous head, neck and breast and mottled black and rufous back. Female usually greyer above. Distinctive bold black bars on the side of belly. In non-breeding plumage, uniform greyish-brown above with white underparts. Juvenile has dark-brown back with buff-coloured wash on neck and breast. In flight, diagnostic broad white wing-bar and white rump and uppertail-coverts contrast with black tail-band.
Voice Sharp *kik-kik*.
Similar species Bar-tailed Godwit has slightly upcurved bill and different wing pattern. See also Asian Dowitcher.

Range Breeds from Europe east to Ussuriland and winters south to Africa and Australia. Common passage migrant in north-eastern China and Korea.
Status in Japan Common passage migrant on coastal mudflats, rice fields and freshwater marshes. Often forms mixed flocks with Bar-tailed Godwit.

Tadao Shimba

Adult ♂ breeding. May, Aichi

Yoshihito Goto

Juvenile. September, Gifu

Tadao Shimba

Juvenile. September, Mie

Bar-tailed Godwit

Limosa lapponica 41cm

Akira Yamamoto

Adult ♂ breeding. May, Aichi

Tadao Shimba

Juvenile. October, Mie

Description Large wader with long, upcurved bill with pale basal half. Female is larger than male. In breeding plumage, male has chestnut face, breast and belly with dark brown mantle and scapulars fringed rufous. Female has dark-brown back and a buff-coloured body. Juvenile similar to female, but overall more buff-coloured. In non-breeding plumage has greyish-brown back and pale greyish-brown body. In flight, rump and uppertail-coverts pale tail heavily barred brown.
Voice Sharp *kerk-kerk*.
Similar species Black-tailed Godwit has longer legs, straight bill and distinctive pied wing and tail pattern.

Range Breeds in northern Eurasia and western Alaska and winters south to Australia and New Zealand. Common passage migrant in coastal north-east Asia.
Status in Japan Common passage migrant on coastal mudflats.

Tadao Shimba

Juvenile. October, Mie

Little Curlew

Numenius minutus 31cm

Description Small curlew, with short, slightly downcurved bill, greyish legs and broad buff-coloured supercilium with dark head stripes. In flight shows buff-coloured rump and tail.
Voice Harsh whistle, often repeated *kweek*.
Similar species Whimbrel is larger with brown-barred white rump and back and has different call.

Tadao Shimba

Adult breeding. May, Hegura-jima

Range Breeds in eastern Siberia and migrates south through eastern Asia. Winters in New Guinea and Australia.
Status in Japan Uncommon migrant on rice fields and grassy areas.

Tadao Shimba

Adult breeding. May, Hegura-jima

Tadao Shimba

Juvenile. October, Aichi

Whimbrel

Numenius phaeopus 41cm

Tadao Shimba

Adult breeding. May, Aichi

Tadao Shimba

Adult breeding. May, Mie

Description Large dark-brown wader with long, dark down-kinked bill, dull greyish legs and striped head pattern. In flight, shows white rump.
Voice Series of loud whistles repeated rapidly *pi-pi-pi-pi*.
Similar species See Bristle-thighed Curlew and Little Curlew.

Range Breeds in northern Eurasia and northern North America and winters further south.
Status in Japan Common passage migrant, especially abundant in spring. Visits a wide range of habitats including rocky shores, coastal mudflats and rice fields. Some winter on Ryukyu Islands.

Bristle-thighed Curlew

Numenius tahitiensis 44.5cm

Kazuyasu Kisaichi

Adult. April, Hawaii, USA

Description Overall resembles Whimbrel, but has mottled dark and cinnamon back and pale cinnamon-coloured underparts. Bristly thighs only visible at very close range. In flight, cinnamon lower rump, upperwing-coverts and tail contrast with brown flight feathers.
Voice Single note *ku-whit,* quite different from that of Whimbrel.

Range Breeds in western Alaska and winters in the tropical Pacific.
Status in Japan Only a small number of records; found on coastal mudflats and grassy fields.

Eurasian Curlew

Numenius arquata 60cm

Description Very large wader with long downcurved bill and dull greyish legs. Brown upperparts, white belly and undertail-coverts. Heavy dark streaks on flanks. In flight shows white back and brown-streaked white rump. Tail barred dark brown. Underwing-coverts white.
Voice Loud, distinctive and repeated *coor-ee* often given in flight.
Similar species Far-eastern Curlew has dark rump and back. Whimbrel is smaller with striped head pattern, pale supercilium and different call.

Akira Yamamoto

Adult. September, Aichi

Range Breeds across Eurasia from Europe east to southern-central Siberia and winters from Europe and Africa east to eastern Asia. Uncommon passage migrant in north-east Asia.
Status in Japan Uncommon passage migrant on coastal mudflats. Common winter visitor in Kyushu; often forms large flocks. Requires extensive mudflats and numbers appear to have declined due to habitat loss.

Akira Yamamoto

Adult. September, Aichi

Tokio Sugiyama

Adult. September, Aichi

Far-eastern Curlew

Numenius madagascariensis 61.5cm

Mitsuo Imai

Adult breeding. April, Chiba

Description Very large wader with long downcurved bill and dull greyish legs. Upperparts dark brown with chestnut feather edging; belly paler with dark streaks. In flight shows dark rump and uppertail-coverts, and brown-barred underwings.

Voice Loud, distinctive and repeated *coor-ee* almost identical to Eurasian Curlew.

Similar species Whimbrel is much smaller with striped head pattern and pale supercilium; has different call. Eurasian Curlew has white back and white underwing-coverts.

Tadao Shimba

Adult non-breeding. December, Queensland, Aus.

Range Breeds in eastern Siberia and north-eastern China and migrates south through eastern Asia. Winters in south-east Asia and Australia.

Status in Japan Uncommon passage migrant to coastal mudflats. One of the first migratory waders to arrive in spring from wintering grounds.

Nobuo Shiraki

Adult breeding. May, Hyogo

Spotted Redshank

Tringa erythropus 32.5cm

Description Medium-sized elegant wader with long red legs, straight dark bill with red base to lower mandible. In breeding plumage has entirely black body with white notches and fringes above and on flanks and undertail-coverts. In non-breeding plumage has white supercilium, grey back and white underparts. Juvenile has greyish-brown neck, breast and belly, and greyish-brown back mottled with white. In flight, plain dark upper-wings contrast with white rump and back.
Voice Loud and rising *chuweet*.
Similar species Common Redshank is smaller, has shorter legs and white secondaries and inner primaries.

Tadao Shimba

Adult breeding. May, Aichi

Range Breeds in northern Eurasia and winters from western Europe, tropical Africa east to south-east Asia. Passage migrant to north-east Asia.
Status in Japan Passage migrant on fresh-water marshes and rice fields. Arrives in early March and remains until early May to complete moult before departing for breeding grounds. Numbers have declined significantly in recent years.

Tadao Shimba

Adult non-breeding. March, Aichi

Tadao Shimba

Juvenile. October, Aichi

Common Redshank

Tringa totanus 27.5cm

Mitsuo Imai

Adult breeding. May, Yonaguni-jima

Description Medium-sized wader with bright red legs and black-tipped red bill. White eye-ring, dark lores and white supercilium that does not extend beyond the eye.
In breeding plumage, back greyish-brown with heavy black streaks. Face, breast and belly whitish, heavily marked brown.
In non-breeding and juvenile plumage, greyish-brown back, face, and breast contrast with white belly. In flight shows white secondaries and inner primaries, white rump and uppertail-coverts.
Voice Loud, ringing *tew-tu-tu*.
Similar species See Spotted Redshank.

Tokio Sugiyama

Juvenile. September, Shizuoka

Range Breeds in Eurasia from Europe east to south-eastern Russia and north-eastern China. Moves south in winter and common passage migrant in Korea.
Status in Japan Uncommon passage migrant to coastal mudflats, rice fields and marshes. Some breed in eastern Hokkaido. Regularly winters on Ryukyu Islands.

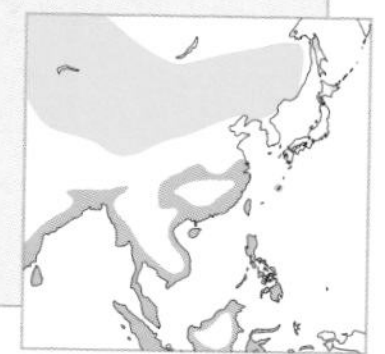

Tadao Shimba

Juvenile. September, Aichi

Marsh Sandpiper

Tringa stagnatilis 24.5cm

Description Elegant, medium-sized wader with long olive-green legs and long needle-like black bill. In breeding plumage brownish-grey above with black spots and bars on upperparts, head and flanks. Non-breeding adults grey above, white below; juveniles are similar but browner above. In flight shows white rump, back and uppertail-coverts.
Voice Loud, high-pitched and often rapidly repeated *yip-yip*.
Similar species Common Greenshank is larger with a stouter and slightly upcurved bill and different call.

Range Breeds from eastern Europe east through central Asia to Lake Baikal. Winters mainly in tropical Africa, southern Asia and Australia. Scarce passage migrant in north-eastern Asia.
Status in Japan Uncommon passage migrant on freshwater marshes and rice fields. Some winter in southern Japan.

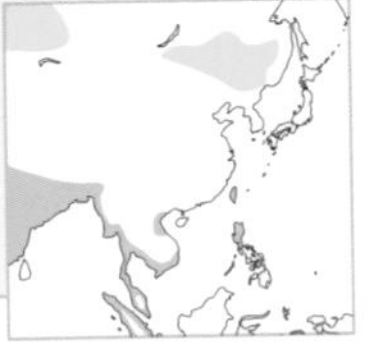

Tadao Shimba

Adult breeding. May, Aichi

Tadao Shimba

Juvenile. September, Okinawa-Hontou

Tadao Shimba

Adult non-breeding. September, Okinawa-Hontou

Common Greenshank

Tringa nebularia 35cm

Tadao Shimba

Adult breeding. April, Ishigaki-jima

Description Medium-large, long-legged wader. Grey upperparts, white underparts, slightly upcurved bill and greyish-green legs. In breeding plumage has heavy black streaks on head, neck and breast; also some dark-centred upperpart feathers. In flight shows white back, rump and uppertail. Normally wary and frequently calls in flight.
Voice Loud, ringing *tew-tew-tew*.
Similar species See Marsh Sandpiper and Nordmann's Greenshank.

Tadao Shimba

Juvenile. September, Mie

Range Breeds across Eurasia east to eastern Siberia and winters from western Europe and Africa east to south-east Asia and Australia. Common passage migrant in north-east Asia.
Status in Japan Common passage migrant in coastal mudflats, estuaries and freshwater marshes. Some winter in southern Japan.

Tadao Shimba

Adult non-breeding. September, Mie

Nordmann's Greenshank

Tringa guttifer 31cm

Description Medium-large wader. Resembles Common Greenshank, but has much shorter yellow legs, pure white underwings and a thicker, straighter bill. Bill dark with pale base. In breeding plumage has black spots on breast. Non-breeding birds have very pale head, pale grey upperparts and white underparts; juveniles similar but grey brown on crown and upperparts. In flight shows white back and rump, and pale grey tail; toes only just project beyond tail-tip.
Voice Distinctive, single note *kehhh*.
Similar species Common Greenshank has a more slender neck, longer greenish legs, brown-barred underwings and a very different call.

Range Breeds on Sakhalin and migrates south through eastern Asia. Winters in southern Asia.
Status in Japan Internationally endangered. Very rare passage migrant on coastal mudflats. Mainly recorded in September.

Tokio Sugiyama

Adult breeding. August, Aichi

Tokio Sugiyama

Adult breeding (in moult). Sept., Aichi

Tadao Shimba

Juvenile. September, Aichi

Greater Yellowlegs

Tringa melanoleuca 35cm

Mitsuo Imai

Adult non-breeding. January, Ibaragi

Description Medium-large, slender wader with long yellow legs. Slightly upcurved dark bill with yellowish-green base. Greyish-brown back with white spots and white underparts. In breeding plumage has heavy black streaks on face and breast and blackish chevrons on underparts. In flight shows square white rump and flight feathers finely spotted white above.
Voice Three or four loud calls; *tu-tu-tu*.
Similar species See Lesser Yellowlegs.

Range Breeds in northern North America; winters in coastal North and South America.
Status in Japan Vagrant. Inhabits coastal mudflats and freshwater marshes. Mainly recorded in summer and autumn.

Lesser Yellowlegs

Tringa flavipes 25cm

Tadao Shimba

Juvenile. August, Ontario, Canada

Tadao Shimba

Juvenile. August, Ontario, Canada

Description Medium-sized, slender wader with long yellow legs and slender, straight dark bill. Greyish-brown back with white spots and white underparts. In breeding plumage, heavily streaked black on face, neck and breast. In flight shows square white rump and uniformly blackish flight feathers above.
Voice Whistled *tu* or *tu-tu,* less musical than Greater Yellowlegs.
Similar species Wood Sandpiper is smaller and call is different. Greater Yellowlegs is larger with a relatively longer, stouter bill.

Range Breeds in northern North America and winters south to South America.
Status in Japan Rare visitor to coastal mudflats, rice fields and freshwater marshes. Some winter records. First recorded in Hokkaido in August 1980.

Green Sandpiper

Tringa ochropus 24cm

Description Medium-sized sandpiper, dark grey-green above with white underparts; distinct eye-ring and white supercilium that does not extend behind the eye. In flight plain dark upperwings and blackish underwings contrast with white underparts, rump and uppertail-coverts. Shy and easily flushed.
Voice High-pitched whistle *twit-twit-twit* or sharp, rising *chew-wit.*
Similar species See Wood Sandpiper.

Norio Yamagata

Adult breeding. May, Aichi

Range Breeds in northern Eurasia and migrates south in winter.
Status in Japan Common passage migrant in freshwater marshes and rivers. Frequently seen along small streams. Many winter in southern Japan.

Tadao Shimba

Juvenile. September, Aichi

Tokio Sugiyama

Juvenile. September, Aichi

Akira Yamamoto

Adult non-breeding. November, Aichi

Wood Sandpiper

Tringa glareola 21.5cm

Tadao Shimba

Adult breeding. May, Aichi

Description Medium-sized sandpiper, greyish-brown above with white underparts. Straight dark bill, dull yellowish legs and clear white supercilium. In breeding plumage has dark streaks on breast and white spots on upperparts. In flight shows conspicuous white uppertail-coverts, rump and white underwings. Normally shy and calls loudly when flushed.
Voice Excited, shrill series of whistled notes *chiff-iff-iff.*
Similar species Green Sandpiper is slightly larger and darker overall; white supercilium does not extend behind the eye and has distinct eye-ring. Green Sandpiper is easily separated by its dark underwing and different call.

Tadao Shimba

Juvenile. September, Aichi

Range Breeds in northern Eurasia and winters in Africa, southern Asia and Australia. Common passage migrant in north-east Asia.
Status in Japan Common passage migrant to rice fields and freshwater marshes. Some winter in southern Japan.

Tadao Shimba

Adult non-breeding. January, Aichi

Terek Sandpiper

Xenus cinerea 23cm

Description Medium-small sized wader with long upcurved bill, greyish back and white underparts. Shortish yellow legs and dark bill with yellow base. In breeding plumage has black mark on scapulars. In flight, white trailing edges to secondaries are visible.
Voice Sharp, fluting *twit-wit-wit-wit.*
Similar species Grey-tailed Tattler is larger and has longer, straight bill.

Tokio Sugiyama

Adult non-breeding. September, Aichi

Range Breeds in northern Eurasia and winters mainly in Africa, south-eastern Asia and Australia. Common passage migrant along coastal areas of north-east Asia.
Status in Japan Common passage migrant throughout Japan on coastal mudflats and estuaries. Very active while feeding and normally wary.

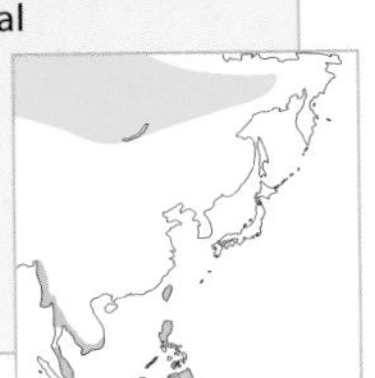

Tadao Shimba

Adult transitional. October, Aichi

Tadao Shimba

Adult non-breeding. September, Ibaragi

Common Sandpiper

Actitis hypoleucos 20cm

Akira Yamamoto

Juvenile. August, Aichi

Description Small, short-legged wader with olive-brown upperparts and white underparts. White shoulders, white supercilium, short dark bill and greenish-grey legs. Constantly bobs body and tail up and down while feeding.
Voice Series of piping notes *tsee-wee-wee.*
Similar species Green Sandpiper is larger, has shorter supercilium from the base of bill to eye, white eye-ring and no wing-bar.

Tadao Shimba

Adult non-breeding. January, Aichi

Range Breeds in temperate and boreal Eurasia and moves south in winter. Common summer visitor to north-east Asia.
Status in Japan Common resident along rivers, lakes and coastal areas in Honshu. Summer migrant to Hokkaido and winters from Honshu southwards.

Spotted Sandpiper

Actitis macularius 19cm

Tadao Shimba

Adult breeding. August, Ontario, Canada

Description Resembles Common Sandpiper but has distinctive black spots on breast and bright orange bill in breeding plumage. Non-breeding plumage almost identical to Common Sandpiper, but has slightly downcurved yellowish bill, unbarred tertials, slightly shorter body and shorter white wing-bar on upperwings.
Voice Similar to Common Sandpiper.

Range Widespread in North America and winters south to South America.
Status in Japan Single record of a bird in breeding plumage on Hokkaido in May 2003.

Grey-tailed Tattler

Heteroscelus brevipes 25cm

Description Medium-sized wader with plain grey upperparts, longish straight bill and short yellow legs. In breeding plumage has heavy grey barring on sides of body.
Voice Clear whistle *tu-whit*.
Similar species In non-breeding plumage may be confused with Red Knot, which has shorter stockier body and a shorter bill. See also Wandering Tattler.

Range Breeds in eastern Siberia and migrates south through coastal areas of eastern Asia. Winters in south-eastern Asia and Australia.
Status in Japan Common passage migrant throughout Japan. Inhabits coastal mudflats and rocky shores. Often the last species among the migratory waders to leave for breeding grounds in late May and returns as early as mid-July. Some winter in southern Japan.

Tadao Shimba

Adult breeding. May, Aichi

Tadao Shimba

Adult breeding. August, Shizuoka

Tadao Shimba

Adult breeding. May, Aichi

Wandering Tattler

Heteroscelus incanus 28cm

Kazuyasu Kisaichi

Adult breeding. May, Chiba

Description Medium-sized wader with plain grey upperparts, longish straight bill and short yellow legs. In breeding plumage, underparts including undertail-coverts are heavily barred. In flight shows faint wing-bar on upperwing.

Voice Whistled series of trilling notes of the same pitch *pew-tu-tu-tu.*

Similar species Grey-tailed Tattler is slightly smaller, paler and has less heavy barring on underparts in breeding plumage. At close range, Wandering's much longer nasal groove is diagnostic. Best separated by call.

Tadao Shimba

Adult non-breeding. December, New Caledonia

Range Breeds in north-eastern Siberia and in Alaska and winters on lower Pacific coast of North America, in the tropical Pacific and eastern Australia.

Status in Japan Rare passage migrant to rocky shores along the Pacific coast of Honshu and Ogasawara Islands. Occurs annually at a small number of sites in mid-May.

Teruaki Ishii

Adult breeding. May, Aichi

Ruddy Turnstone

Arenaria interpres 22cm

Description Medium-sized stocky wader with distinctive face and breast pattern, short, slightly upturned black bill, and short orange legs. In breeding plumage, rufous with black on back contrasting with white underparts. In non-breeding plumage lacks rufous on back and has dull greyish-brown head pattern.
Voice Clear, rapid, rolling *trik-tuk-tuk-tuk*.

Tadao Shimba

Adult breeding. May, Mie

Range Breeds across arctic Eurasia and North America; disperses to warmer coastal areas in winter.
Status in Japan Common passage migrant on coastal mudflats and rocky shores. Some may winter in southern Japan.

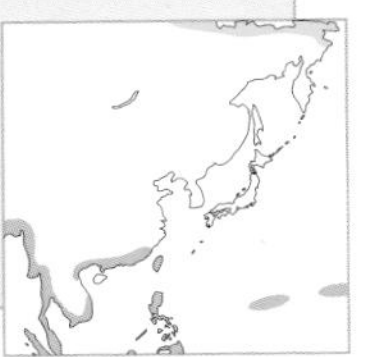

Tadao Shimba

Adult breeding. May, Aichi

Tadao Shimba

Adult non-breeding. December, Queensland, Australia

Great Knot

Calidris tenuirostris 28.5cm

Mitsuo Imai

Adult breeding. May, Yonaguni-jima

Description Stocky wader with straight black bill and dark legs. In breeding plumage has broad black band on breast, black-spotted underparts and distinctive rufous spots on scapulars. In non-breeding plumage has uniform light grey back and white underparts, with dark streaks on breast and flanks. Juvenile has heavy black streaks on breast and dark feathers on back have white edges. In flight shows clear white wing-bar and white uppertail-coverts.
Voice Not very vocal; double note whistle *nyut-nyut.*
Similar species See Red Knot.

Tadao Shimba

Juvenile. September, Mie

Range Breeds in north-east Siberia and winters in south-eastern Asia and Australia. Common passage migrant in north-east Asia.
Status in Japan Common passage migrant on coastal mudflats and sandy shores. Rare on inland freshwater habitats. Often flocks with Bar-tailed Godwit.

Tadao Shimba

First-winter. October, Mie

Red Knot

Calidris canutus 24.5cm

Description Overall shape similar to Great Knot, but smaller with shorter bill and yellowish legs. In breeding plumage has rufous face and breast, and dark back mottled with rufous and black spots. In non-breeding plumage pale-grey above with white underparts. Juvenile has faint streaks on breast and dark subterminal markings on upperparts, giving a 'scaly' appearance. In flight shows narrow white wing-bar on upperwings and pale grey rump.
Voice Not very vocal; loud and slightly harsh *knut*.
Similar species Grey-tailed Tattler has slimmer body and longer bill.

Range Breeds in high arctic and winters in tropical areas and southern hemisphere. Passage migrant in north-east Asia.
Status in Japan Uncommon passage migrant to coastal mudflats and sandy shores. Rare in spring, often seen in autumn among flocks of Great Knot.

Tadao Shimba

Juvenile (right, with juvenile Great Knot left). September, Mie

Tadao Shimba

Adult breeding. May, Mie

Tadao Shimba

Juvenile. September, Mie

Sanderling

Calidris alba 19cm

Tadao Shimba

Adult breeding. May, Aichi

Description Small stocky wader with short black bill and black legs. In breeding plumage has rufous head, breast and mottled rufous and black back. In non-breeding plumage plain grey above with white underparts. Juvenile has mottled black and white back with distinctive black shoulder-patch. In flight shows broad white wing-bar on upperwings. Very active when feeding.

Voice Sharp, short and whistled *twick-twick.*

Similar species Non-breeding Dunlin is about the same size and is similar in colour, but has a longer and downcurved bill.

Tadao Shimba

Juvenile. October, Mie

Range Breeds in high arctic and winters in coastal areas of warmer regions. Passage migrant in north-east Asia.

Status in Japan
Common passage migrant and winter visitor on sandy coasts. Often seen in flocks.

Tadao Shimba

Non-breeding. October, Mie

Western Sandpiper

Calidris mauri 16cm

Description Small wader with long, slightly downcurved bill and black legs. In breeding plumage shows rufous markings on head, ear-coverts and upperparts, and fine black streaks on neck and breast. Juveniles have blackish mantle feathers, fringed rufous. In non-breeding plumage has uniform grey back contrasting with white underparts. **Similar species** Dunlin is larger and Broad-billed Sandpiper has distinctive split supercilium. Other small stints have shorter bills.

Norio Yamagata

Adult breeding. April, Shizuoka

Tadao Shimba

Adult non-breeding. January, Florida

Range Breeds on Chukotski peninsula in north-eastern Russia and in western Alaska. Winters south to northern South America. **Status in Japan** Mainly autumn vagrant. Inhabits coastal mudflats and shallow waters.

Baird's Sandpiper

Calidris bairdii 19cm

Description Small wader with dark brown back, fine streaks on breast and white belly. Short black straight bill and short black legs. At rest, primaries project well beyond tail tip. In flight shows inconspicuous white wing-bar across upperwings and white edges to rump and uppertail-coverts. **Similar species** This species can be separated from other small waders by its long primaries and horizontal posture.

Range Breeds on Chukotski Peninsula in North-eastern Russia and in arctic North America. Winters in South America. **Status in Japan** Vagrant. Coastal mudflats and rice fields.

Akira Yamamoto

Juvenile. September, Aichi

Red-necked Stint

Calidris ruficollis 15cm

Tadao Shimba

Adult breeding. May, Aichi

Tadao Shimba

Juvenile. September, Aichi

Description Small wader with short black bill and short black legs. In breeding plumage has rufous face, head and breast; mantle and scapular feathers rufous and black with white tips, grey upperwing-coverts and white underparts. In non-breeding plumage is greyish above and white below. Juvenile has brown-streaked head, white supercilium, and mottled brown and black back and grey upperwing-coverts. In flight shows narrow white wing-bar, a dark brown centre to rump and tail, white rump sides and grey outer tail feathers.
Voice High-pitched *thrrrh.*
Similar species Long-toed Stint is browner, has long yellow legs. See also Little Stint.

Range Breeds in arctic Siberia. Winters in south-east Asia, Australia and New Zealand. Common passage migrant in north-east Asia.
Status in Japan The most common small wader but numbers apparently declining. Passage migrant on coastal mudflats, rice fields and freshwater marshes. Some winter in southern Japan.

Akira Yamamoto

Adult non-breeding. September, Okinawa-Hontou

Little Stint

Calidris minuta 13.5cm

Description In breeding plumage has rufous head, neck and breast washed with black streaks, white face and throat, mantle and scapular feathers and upperwing coverts rufous and black with white tips, and white underparts. In non-breeding plumage, greyish above and white below. Juvenile has mottled brown and black back; white tips of upperpart feathers form white 'V' lines above.
Voice Thin *chit.*
Similar species Red-necked Stint almost identical in winter but longer-bodied and shorter-legged; otherwise has rufous throat and upper-breast. Western Sandpiper has longer, curved bill.

Range Breeds in northern Scandinavia and arctic Siberia. Winters mainly in southern Europe, Africa, the Middle East and the Indian subcontinent.
Status in Japan Rare passage migrant on coastal mudflats and sandy shores. Some winter records.

Kazuyasu Kisaichi

Adult breeding. April, Chiba

Kazuyasu Kisaichi

Juvenile. July, Ibaragi

Akira Yamamoto

Adult non-breeding. January, Aichi

Temminck's Stint

Calidris temminckii 14.5cm

Tadao Shimba
Adult breeding. May, Aichi

Description Small wader with short black bill and yellowish legs. In breeding plumage plain grey-brown above with fine black streaks on face and breast and mottled black and rufous on back. In non-breeding plumage, uniform dark grey head, face, breast and back contrasting with white underparts. Juvenile is similar to non-breeding adult, but has pale fringes and dark subterminal lines on scapulars and upperwing-coverts.
Voice Metallic, soft *chirri-rrih.*
Similar species Red-necked Stint has black legs and short neck. Long-toed Stint is browner, has longer legs and neck and a pale supercilium.

Tadao Shimba
First-winter. March, Aichi

Range Breeds across northern Eurasia and winters mainly in Africa, Mediterranean and southern Asia. Uncommon passage migrant in north-east Asia.
Status in Japan Uncommon passage migrant on freshwater marshes and rice fields. Regularly winters from mid-Honshu southwards.

Tadao Shimba
Adult non-breeding. April, Aichi

Long-toed Stint

Calidris subminuta 15cm

Description Small wader with relatively long yellow legs. Tends to assume more upright posture than other stints. In breeding plumage has a brown head and buff-coloured breast, finely streaked black; black mantle and scapular feathers edged with brown and tipped white. In juvenile plumage, white edges to upperpart feathers form a V-shaped line on mantle. In non-breeding plumage has dark grey upperparts and white underparts. Shows broad white supercilium in all plumages.
Voice Soft, short *prrp-prrp*.
Similar species Little Stint and Red-necked Stint have black legs and shorter neck. Temminck's Stint has shorter legs, shorter neck and more horizontal posture. Sharp-tailed Sandpiper is larger.

Tadao Shimba

Adult breeding. May, Aichi

Range Breeds sparsely in eastern Siberia and the northern Kurile Islands. Winters in south-east Asia and Australia. Uncommon passage migrant to north-east Asia.
Status in Japan Uncommon passage migrant on freshwater marshes and rice fields. Most common in autumn. Some winter in southern Japan.

Mitsuo Imai

Juvenile. August, Ibaragi

Tadao Shimba

Adult non-breeding. March, Aichi

Pectoral Sandpiper

Calidris melanotos 22cm

Tadao Shimba

Adult non-breeding. August, Ontario, Canada

Description Very similar to Sharp-tailed Sandpiper but darker overall, with longer neck and darker bill with pale base. Uniform dark streaks on breast contrast sharply with white belly. In flight shows white wing-bar and white sides to rump and uppertail-coverts.

Tadao Shimba

Juvenile. September, Ontario, Canada

Range Breeds in north-eastern Russia and arctic North America. Winters mainly in South America, also Australia and New Zealand.

Status in Japan Rare summer/autumn visitor to rice fields, freshwater marshes and coastal mudflats.

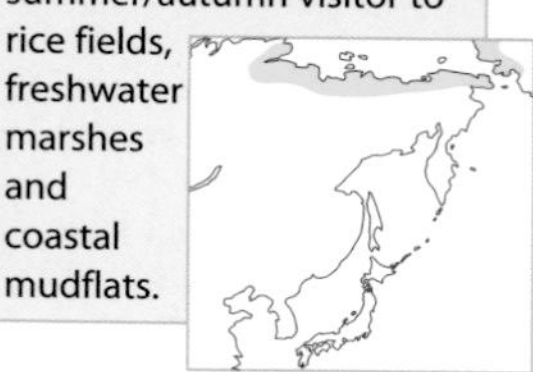

Tsunehiro Komai

Adult breeding (in moult). August, Aichi

Sharp-tailed Sandpiper

Calidris acuminata 21.5cm

Description Medium-sized wader with slightly downcurved bill and yellow legs. In breeding plumage, rufous cap, white supercilium, dark-brown upperpart feathers fringed chestnut and white; rufous-brown breast and white belly, streaked boldly. Juvenile has bright rufous-buff on breast with faint short black streaks on neck. Richly-coloured upperparts and white edges to mantle feathers make V-shape pattern on back. Non-breeding adult is greyer with fine streaks on neck, breast and flanks. In flight shows white wing-bar and white uppertail-coverts.
Voice Soft *pleep-pleep*.
Similar species See Pectoral Sandpiper.

Tadao Shimba

Adult ♀ breeding. May, Aichi

Tadao Shimba

Adult non-breeding. December, Queensland

Range Breeds in northern Siberia and winters mainly in Australasia.
Status in Japan Passage migrant to marshes, rice fields and coastal mudflats of north-east Japan. Common in spring, less frequent in autumn.

Tadao Shimba

Juvenile. September, Aichi

Curlew Sandpiper

Calidris ferruginea 21.5cm

Tokio Sugiyama

Adult ♂ breeding. May, Tsushima

Description Medium-small wader with long downcurved black bill and black legs. In breeding plumage unmistakable, with brick-red head, face and underparts, and upperparts richly marked with black, white and chestnut. Juvenile has grey-brown upperparts with pale feather-edges and fine streaks on breast. In non-breeding plumage, greyish above with white underparts. In flight shows white wing-bar across upperwings and distinctive white rump and uppertail-coverts.
Voice High-pitched, stuttering *trrrup.*
Similar species See Dunlin.

Range Breeds in northern Siberia and winters in tropical Africa, southern Asia and Australia. Uncommon passage migrant in north-east Asia.
Status in Japan Uncommon passage migrant on freshwater marshes, coastal mudflats and rice fields.

Akira Yamamoto

Juvenile. September, Aichi

Tadao Shimba

Adult non-breeding. December, Queensland, Australia

Dunlin

Calidris alpina 21cm

Description Medium-small wader with long downcurved bill and black legs. In breeding plumage has black streaks on breast, distinctive large black patch on belly and rufous upperparts with black spots. Juvenile has spotted flanks with white 'V's on mantle and scapulars. In non-breeding plumage, plain grey above and white below. In flight shows white wing-bar and white sides to rump and uppertail-coverts.
Voice Distinctive, nasal, trilled *jureeep.*
Similar species Juvenile and non-breeding plumaged Curlew Sandpipers show longer bill and neck and distinct white rump in flight.

Range Breeds across arctic Eurasia and North America. Moves south in winter and common winter visitor to Korea.
Status in Japan Common winter visitor to coastal mudflats. Often forms large flocks. Numbers have declined since the 1980s.

Tadao Shimba

Adult breeding. May, Aichi

Akira Yamamoto

Adult non-breeding. November, Aichi

Tadao Shimba

Adults, breeding. May, Aichi

Rock Sandpiper

Calidris ptilocnemis 21cm

Tokio Sugiyama

Adult non-breeding. March, Chiba

Description Medium-small wader with slightly downcurved bill with yellow base, and short yellow legs. In breeding plumage has a large black patch on lower breast, white face with dark spot on cheek and rufous edges to mantle and scapulars. Non-breeders are dark grey above, paler below, with streaked neck and breast. In flight shows white wing-bar and white outer tail feathers.
Similar species Non-breeding Dunlin has longer bill, black legs and paler grey back.

Range Breeds in north-eastern Russia, western Alaska and on the Aleutian Islands. Winters on the northern coasts of the Pacific Ocean.
Status in Japan Rare winter visitor to rocky shores of northern Japan.

Stilt Sandpiper

Calidris himantopus 22cm

Tokio Sugiyama

Adult breeding. August, Tokyo

Description Medium-sized wader with distinctive long downcurved black bill and long yellow legs. Non-breeders are greyish above with white underparts streaked grey, and a white supercilium. In breeding plumage shows black spots on back, rust-coloured ear-coverts and heavy barring on breast and belly. In flight shows white uppertail-coverts.
Similar species Curlew Sandpiper has short black legs.

Kazuyasu Kisaichi

Adult breeding (in moult). August, Ibaragi

Range Breeds in northern North America and winters mainly in South America.
Status in Japan Vagrant. Four records from Honshu and first recorded in Aichi in July 1977. Inhabits freshwater marshes and coastal mudflats.

Spoon-billed Sandpiper

Eurynorhynchus pygmeus 15cm

Description Small wader with unique, spoon-shaped bill and short black legs. In breeding plumage has rufous head, face and breast; black feathers of mantle and scapulars edged rufous, and white underparts. Most frequently seen juvenile plumage has dark crown, white supercilium and dark upperpart feathers edged white. In non-breeding plumage, uniform grey above contrasting with white underparts.
Voice Soft *puri-riih.*
Similar species Juvenile Sanderling has similar colour pattern and may be confused from a distance.

Range Breeds on Chukotski Peninsula and migrates south through eastern Asia. Winters in south-east Asia.
Status in Japan Internationally endangered. Rare but annual passage migrant in coastal areas. Prefers sandy habitats. Mainly seen in autumn and there are a few winter records.

Tokio Sugiyama

Juvenile. September, Aichi

Tadao Shimba

Juvenile. September, Aichi

Tadao Shimba

Broad-billed Sandpiper

Limicola falcinellus 17cm

Tadao Shimba

Adult breeding. May, Aichi

Akira Yamamoto

Juvenile. September, Aichi

Description Small wader with distinctive long downcurved bill with heavy base, black legs and split white supercilium. In breeding plumage, heavily streaked on breast; white edges to mantle and scapulars form 'V's. In non-breeding plumage has faint black streaks on breast and pale grey back. Juvenile has upperpart back feathers edged with rufous and white. In flight shows white wing-bar and white outer uppertail-coverts.
Voice Trilled *chrrect,* similar to Dunlin.
Similar species Dunlin is larger and does not have split supercilium.

Range Breeds in northern Scandinavia and Siberia and winters from east Africa to Australia. Passage migrant to north-east Asia.
Status in Japan Passage migrant on coastal mudflats. Rare in spring and uncommon in autumn. Juveniles are most often seen.

Akira Yamamoto

Adult non-breeding. October, Aichi

Ruff

Philomachus pugnax ♂32cm ♀25cm

Description Distinctive medium-large wader, with small head, long neck; short, slightly downcurved bill and yellowish-red legs. Male in breeding plumage has colourful ruff and ear-tufts that vary greatly in colour but mainly reddish-coloured birds occur in Japan. Female (reeve) is much smaller than male. In breeding plumage, barred upperparts and breast contrasts with white underparts. Non-breeding plumage of both sexes has uniform grey back with white underparts. Juvenile has buff-coloured breast, white belly and dark-brown back. Buff edges to dark upperpart feathers give a scaly appearance. In flight shows narrow wing-bar and oval-shaped white sides to uppertail-coverts.
Voice Not a vocal species.
Similar species Sharp-tailed Sandpiper may be confused with female, but is smaller, has shorter legs and clearer white supercilium.

Range Breeds across northern Eurasia and winters mainly in Africa, southern Europe, the Middle East and the Indian subcontinent.
Status in Japan
Uncommon passage migrant on freshwater marshes and rice fields. Some winter locally in southern Japan.

Kazuyasu Kisaichi

Adult ♂ breeding. May, Finland

Norio Yamagata

Adult ♂ breeding. May, Aichi

Tadao Shimba

Adult ♀ non-breeding. May, Aichi

Tokio Sugiyama

Adult ♂ non-breeding, December, Aichi

Buff-breasted Sandpiper

Tryngites subruficollis 20cm

Kenji Takehara

Juvenile. November, Okinawa-Hontou

Description Medium-small wader with buff-coloured face, breast and belly, scaly brown and buff upperparts, obvious dark eye, short black bill and yellow legs. White eye-ring and no obvious supercilium. In flight shows white underwings.
Similar species Pacific Golden Plover inhabits same habitat, but is larger, has shorter bill and black legs.

Range Breeds mainly in arctic North America and winters in southern South America.
Status in Japan Vagrant. Occurs mainly around rice fields and occasionally coastal mudflats. Also recorded from Korea.

Wilson's Phalarope

Phalaropus tricolor 23cm

Tadao Shimba

Adult ♀ breeding. May, Ontario, Canada

Tadao Shimba

Adults breeding. June, Ontario, Canada

Description Medium-sized wader with long needle-shaped bill. In breeding plumage, female has unmistakable chestnut, grey and black upperparts and black legs. Breeding male is duller and browner. In non-breeding plumage has grey back and white underparts with yellow legs. In flight, white uppertail-coverts are visible.
Voice Not a vocal species, low *work* or *chek*.
Similar species Marsh Sandpiper has longer, greenish legs. In winter other phalaropes have shorter bills and black eye-patches.

Range Breeds in prairie regions of North America and winters in southern South America.
Status in Japan Vagrant. Two records in Aichi in May 1985 (male) and July 1986 (female). Also recorded once in Korea.

Red-necked Phalarope

Phalaropus lobatus 19cm

Description Distinctive small wader, with small head, needle-like bill and short dark legs. In breeding plumage, dark-brown upperparts with buff-coloured lines on mantle and scapulars, white chin and throat with diagnostic red neck. Female is brighter than male. Juvenile has dark crown, white face, conspicuous black eye-patch and dark-brown upperparts fringed buff. In non-breeding plumage has uniform grey crown and upperparts and a dark eye-patch.
Voice Soft and repeated *chek*.
Similar species Grey Phalarope is larger, has thick neck and bill.

Range Breeds in northern Eurasia and North America. Winters mainly in Arabian Sea and Indian Ocean.
Status in Japan Common passage migrant. Mainly pelagic, but flocks may visit coastal areas in rough seas. Occasionally on inland waters.

Teruaki Ishii

Adults breeding. May, Hegura-jima

Tokio Sugiyama

Juvenile. September, Aichi

Akira Yamamoto

Juvenile. September, Aichi

Grey Phalarope

Phalaropus fulicarius 22cm

Kazuyasu Kisaichi

Adult in moult. July, Ibaragi

Other name Red Phalarope

Description In breeding plumage has reddish-brown neck and underparts; dark cap with distinctive large white patch on face, dark upperpart feathers tinged buff, and yellow bill. Female brighter than male. In non-breeding plumage has uniform grey back, white face with dark eye-patch, black bill with yellow at base.

Voice Low *whit.*

Similar species Non-breeding Red-necked Phalarope is similar but smaller with a needle-like bill.

Tokio Sugiyama

Adult non-breeding. January, Aichi

Range Breeds in high arctic in Eurasia and North America. Winters mainly at sea off western Africa and western South America.

Status in Japan Uncommon passage migrant, mostly pelagic. Rare in coastal areas under rough sea conditions.

Kazuyasu Kisaichi

Adult ♀ breeding. July, Ibaragi

South Polar Skua

Catharacta maccormicki 59cm

Description Large stocky brown seabird with hooked bill and broad wings with crescent-shaped white flashes at base of primaries above and below. Short wedge-shaped tail with two central feathers slightly projecting. Wings are broader and more rounded than in other skuas. Flies steadily with slow wingbeats and little gliding. Piratical; steals food from other seabirds.
Similar species Separated from immature large gulls by large white patches at base of primaries. Other skuas are smaller and slimmer.

Mike Carter

Adult. February, Victoria, Australia

Range Breeds in Antarctic and disperses to northern oceans after breeding.
Status in Japan Uncommon passage migrant in pelagic waters in Honshu and Hokkaido. Most often observed in late spring and early summer.

Pomarine Skua

Stercorarius pomarinus 51cm

Description Medium-sized skua with distinctive paddle-shaped, twisted central tail feathers and white patch at base of outer primaries. Two colour morphs; light morph has dark-brown cap with yellow collar and white underparts; dark morph uniform dark brown. In non-breeding plumage, dark barring on throat, flanks and undertail-coverts, and shorter tail. Juvenile is mottled dark-brown and white overall with heavily barred underparts and underwings. Flies steadily with slow wing beats and little gliding. Piratical; steals food from other seabirds.
Similar species Arctic Skua is smaller, has slimmer body and narrower wings.

Kazuyasu Kisaichi

Adult non-breeding. January, Chiba

Range Breeds in Arctic and disperses south to warmer oceans in winter.
Status in Japan Uncommon passage migrant in offshore waters around Honshu and Hokkaido. Occasionally visits coastal waters following flocks of Common Terns in spring. Some remain in winter.

Arctic Skua

Stercorarius parasiticus 46cm

Akira Yamamoto

Adult breeding. May, Aichi

Description Medium-small skua with pointed central tail feathers and white flashes at primary bases. Two colour morphs; light morph has dark-brown cap with yellow collar and white underparts; dark morph uniform dark-brown. In non-breeding plumage has dark barring on throat, flanks and undertail-coverts, and shorter tail. Juvenile is mottled above and has barred belly. Flies steadily with slow wingbeats and little gliding. Piratical; steals food from other seabirds. **Similar species** Pomarine Skua is more heavily built. Long-tailed Skua is slimmer with greyer upperparts.

Range Breeds mainly in Arctic and disperses south to warmer oceans in winter.
Status in Japan Uncommon passage migrant in offshore waters around Honshu and Hokkaido. Less numerous than Pomarine Skua.

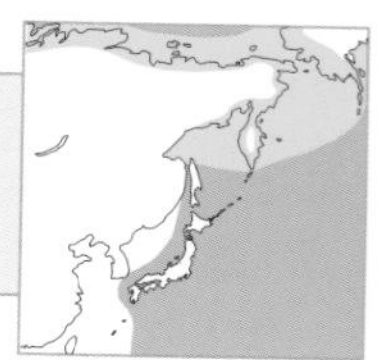

Long-tailed Skua

Stercorarius longicaudus 53cm

Kazuyasu Kisaichi

Adult breeding. July, Hokkaido

Description Smallest skua, with distinctive, very long, central-tail feathers lacks white wing-flashes. Two colour morphs; light morph has black cap with yellow collar and white underparts; extremely rare dark morph uniform dark-brown. In non-breeding plumage has dark barring on throat and flanks and shorter tail. Juvenile has mottled dark and white back with heavily barred underparts. Flight is more buoyant than other skuas.
Similar species Both Pomarine and Arctic Skua are larger and more heavily built with white patches at base of outer primaries.

Range Breeds in high arctic in summer and disperses south to warmer oceans in winter.
Status in Japan Uncommon passage migrant in offshore waters of Honshu and Hokkaido. Most frequently seen off Pacific coast of Honshu in spring.

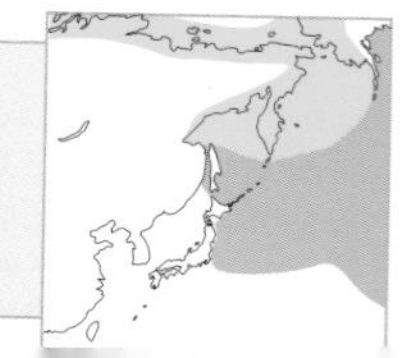

Black-tailed Gull

Larus crassirostris 47cm WS120cm

Description Medium-sized gull. Adult has broad black subterminal tail-band; dark-grey upperwing with black primaries and white trailing-edges to secondaries, a long red-tipped yellow bill with black band, and yellow legs. Head and nape become greyer in non-breeding plumage. Adult plumage is attained after four years. Juvenile is dark-brown with buff-coloured edges to mantle and scapulars, pale forehead, throat and belly and has pinkish bill with dark tip.
Similar species Juvenile Common Gull is smaller and greyer and has a shorter, slimmer bill.

Tadao Shimba

Adult breeding. May, Hegura-jima

Range Breeds from Sakhalin and Kurile Islands south to south-eastern China and winters south to southern China.
Status in Japan Abundant resident in coastal waters from Kyushu northwards. Breeds in colonies on rocky shores and winters mainly from central Honshu southwards. Rare on Ryukyu Islands.

Tadao Shimba

Adult non-breeding. October, Mie

Tadao Shimba

Juvenile. August, Aichi

Tadao Shimba

First-summer. August, Aichi

Common Gull

Larus canus 45cm WS115cm

Tadao Shimba

Adult non-breeding. February, Aichi

Other name Mew Gull

Description Medium-sized gull. Adult has grey upperwings, black outer primaries with white spots near the tips of upperwings, yellowish-green bill and yellow legs. Fine brown streaks on head and breast in non-breeding plumage, when some individuals have a black band on bill. Adult plumage is attained after three years. Juvenile has dark bill, pinkish legs, greyish-brown mantle and black subterminal tail-band.

Similar species Black-tailed Gull has darker upperwings and longer, heavier bill. Vega Gull is much larger and has larger bill with red spot on lower mandible.

Tadao Shimba

Adult non-breeding. February, Aichi

Range Breeds in northern Eurasia and in Alaska. Winters south to the Mediterranean, Persian Gulf, eastern Asia and Pacific coast of North America. In north-east Asia, breeds on lakes from Sakhalin northward and winters south to Korea and Japan.

Status in Japan Common winter visitor to coastal waters from Kyushu northwards.

Tadao Shimba

First-winter. December, Hokkaido

Glaucous-winged Gull

Larus glaucescens 64cm WS135cm

Description Large, bulky, short-winged gull with dark eyes. Adult has grey upperwings, yellow bill with red spot on lower mandible and pink legs. Adult non-breeding has greyish-brown streaks on head and breast. Adult plumage is attained after four years. First–winter is mottled greyish-brown overall with buff-coloured primaries and dark bill.
Similar species First–winter is similar to other large gulls; first–winter Slaty-backed Gull has whiter head with mottled brown and white back and darker tail; first–winter Vega Gull has dark-brown primaries contrasting with pale-brown forewings.

Range Breeds on cliffs around coastal areas of Bering Sea from Kamchatka east to Alaska and northern Canada. Winters south to temperate regions.
Status in Japan Common winter visitor to coastal waters in northern Japan, uncommon elsewhere.

Tadao Shimba

Adult non-breeding. December, Hokkaido

Tadao Shimba

Adult non-breeding. December, Hokkaido

Tadao Shimba

First-winter. December, Hokkaido

Glaucous Gull

Larus hyperboreus 71cm WS135cm

Tadao Shimba

Adult non-breeding. January, Hokkaido

Description A very large, pale gull. Adult has pale-grey upperwings and white primaries, yellow bill with red spot on lower mandible and pink legs. Adult non-breeding has heavy greyish-brown streaks on white head and breast. Adult plumage is attained after four years. First–winter shows long, pink, dark-tipped bill and mottled pale-brown and white upperwings with white primaries.
Similar species Immature Glaucous-winged Gull has buff-coloured primaries and black bill.

Tadao Shimba

Adult non-breeding. January, Hokkaido

Range Breeds in Arctic and moves south in winter. In north-east Asia, breeds on northern shores and winters in coastal areas south to Japan and Korea.
Status in Japan Common winter visitor to coastal areas in northern Japan, uncommon in the south.

Tadao Shimba

First-winter. January, Hokkaido

Thayer's Gull

Larus thayeri 61cm WS140cm

Description Adult is similar to Vega Gull but slightly smaller, has dark eyes, rounded head, shorter bill and legs; grey upperwings with dark-grey outer primaries and white underside to primaries. Adult non-breeding has greyish-brown markings on the white head and neck. Adult plumage is attained after four years. First-winter plumage uniformly greyish-brown overall with white underside to primaries.

Tokio Sugiyama

Adult non-breeding. December, Aichi

Tsunehiro Komai

Adult non-breeding. December, Aichi

Range Breeds in arctic Canada and winters on Pacific coast of North America.
Status in Japan Rare winter visitor to northern Japan and a few regularly winter at a few sites along the Pacific coast of Honshu.

Kazuyasu Kisaichi

First-winter. March, Chiba

Slaty-backed Gull

Larus schistisagus 65cm WS135cm

Tadao Shimba

Adult non-breeding. December, Hokkaido

Tadao Shimba

Adult non-breeding. December, Hokkaido

Description Adult is a large gull with dark-grey upperwings and black outer primaries with white spots at tips, broad white trailing-edges to secondaries and tertials, white underwings with grey primaries, yellow bill with red spot on lower mandible and pink legs. Adult non-breeding has fine greyish streaks on the white head and neck. Adult plumage is attained after four years. First–winter has dark bill and greyish-brown upperwings with brown primaries and dark trailing-edges. Second–winters have a pale bill with dark subterminal bar.

Similar species First–winter Vega Gull has brown head and dark primaries.

Range Breeds from Kamchatka Peninsula south to northern Japan. Some move south in winter.

Status in Japan Common breeding resident on Hokkaido and northern Honshu. Common winter visitor to coastal areas and bays of central Honshu. Uncommon in southern Japan.

Tadao Shimba

Second–winter. December, Hokkaido

Heuglin's Gull

Larus heuglini 58cm WS126–155cm

Description Adult is a large gull with dark slaty-grey upperwings, black primaries, a small rounded-head, yellow bill with large red spot on lower mandible and yellow legs. Adult plumage attained after three years. First–winter has a grey-brown back with pale fringes to back feathers giving a more scaly appearance than other immature gulls.
Similar species Adult Vega Gull is slightly larger, has light-grey upperwings and pinkish legs.

Tsunehiro Komai

Adult non-breeding. January, Aichi

Range Breeds on northern shores of Siberia from Kola Peninsula east to Taimyr Peninsula and winters south to Arabian Peninsula, the Indian subcontinent and south-eastern China. Rare winter visitor to Korea.
Status in Japan Rare winter visitor to coastal areas. Most often seen in western Japan.

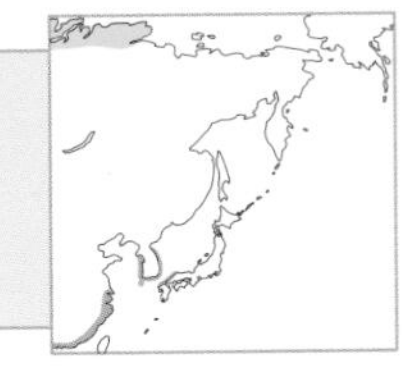

Pallas's Gull

Larus ichthyaetus 69cm WS126–155cm

Description Adult is a large gull with yellow bill with black band and red tip, and yellow legs; grey upperwings with white outer primaries with subterminal black bar. Adult breeding has black hood with white eye-ring; white head with dark patch on ear-coverts in non-breeding plumage. Adult plumage is attained after three or four years. First–winter has white head with dark patch on ear-coverts, greyish-brown upperwings and black subterminal tail-band and a pink bill with black band.

Yoshihito Goto

Adult non-breeding. December, Kumamoto

Range Breeds locally from the Black Sea east to Mongolia. Winters mainly from the eastern Mediterranean to the Indian subcontinent. In north-east Asia, rare winter visitor to Korea and southern Japan.
Status in Japan Rare winter visitor to Kyushu; small numbers regularly winter in Kuma River in Kumamoto.

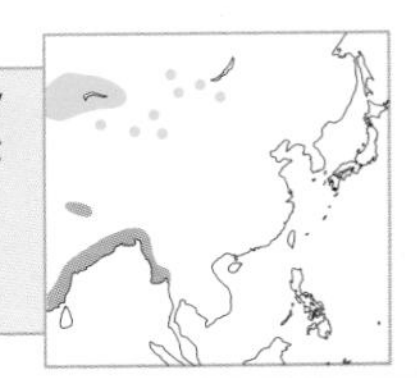

Vega Gull

Larus vegae 62cm WS135cm

Tadao Shimba

Adult breeding. March, Mie

Description Adult is a large gull with grey upperwings, black outer primaries with two white mirrors; yellow bill with red spot to lower mandible and pink legs. Adult non-breeding has fine greyish-brown streaks on white head and breast.
Adult plumage is attained after four years. First–winter is pale-brown with dark outer primaries, pale inner primaries and dark bill with pale basal half.
Similar species First–winter Slaty-backed Gull has greyer upperparts and browner primaries. See also Thayer's Gull, Mongolian Gull and Heuglin's Gull.

Tadao Shimba

Adult breeding. March, Mie

Range Breeds in north-eastern Siberia and winters from Kamchatka Peninsula south to eastern China.
Status in Japan Common winter visitor to coastal waters and large lakes.

Tadao Shimba

First-winter. March, Mie

Mongolian Gull

Larus mongolicus 62cm WS128–158cm

Description Adult is similar to Vega Gull, but head flatter and has smaller red spot on lower mandible and slightly longer, pinkish legs; head less streaked than non-breeding Vega Gull. First–winter has white head, pale back with brown spots and a narrow black subterminal tail-band.
Similar species First–winter separated from Vega Gull by longer head, whiter back and narrower subterminal tail-band.

Hyun-tae Kim

Adult breeding. April, Sosan, South Korea

Range Breeds from eastern Altai east to Lake Baikal and Mongolia. Winters in Korea, western Japan and eastern China.
Status in Japan Rare winter visitor to coastal areas. Most often seen in western Japan.

Hyun-tae Kim

Adult breeding. April, Sosan, South Korea

Hyun-tae Kim

First-winter. December, Sosan, South Korea.

Black-headed Gull

Larus ridibundus 40cm WS93cm

Tadao Shimba

Adult breeding. April, Mie

Tadao Shimba

Adult non-breeding. January, Aichi

Description Medium-small gull with pale-grey upperwings, red bill and legs. White wedge on outer primaries and black trailing-edge to primaries and secondaries. Middle primaries dark grey below. Adult breeding has chocolate-brown hood with white eye-ring; in non-breeding plumage has white head with dark patch on ear-coverts. Juvenile has orange bill with dark tip, orange legs, mottled brown bar on upperwing-coverts, broad brown trailing-edges to upperwings and black subterminal tail-band.

Similar species Saunders's Gull is smaller, has stouter black bill and white outer primaries of underwing contrasting with inner black primaries. Bonaparte's Gull is smaller, has black bill and white underside to primaries.

Range Widely distributed in Eurasia and moves to warmer regions in winter.

Status in Japan Abundant winter visitor to both coastal and inland waters from Kyushu northwards. Rare on Ryukyu Islands. Often tame in city parks in winter.

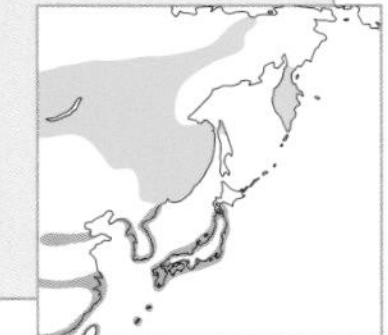

Tadao Shimba

First-winter, December, Aichi

Slender-billed Gull

Larus genei 42–44cm WS105cm

Description Similar to Black-headed Gull but slightly larger, and has longer neck, flatter head and long dark-reddish bill. White underparts with breast and belly tinged pink in breeding plumage.
Similar species Separated from other small gulls by distinctive long bill and neck, and pale iris.

Teruaki Ishii

Adult non-breeding. February, Fukuoka

Range Breeds from the Mediterranean east to central Asia and winters mainly in the Mediterranean, Persian Gulf and Red Sea.
Status in Japan Vagrant. One or two individuals regularly wintered in northern Kyushu among flocks of Black-headed Gulls from 1984 to 1991. Occurs at river mouths and coastal mudflats.

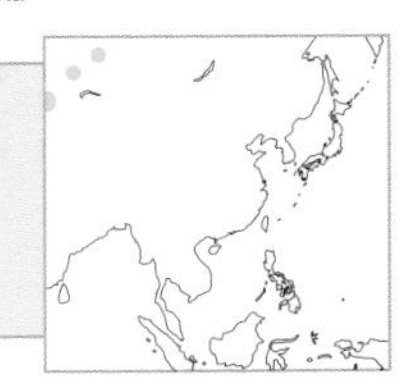

Bonaparte's Gull

Larus philadelphia 36cm WS82cm

Description Similar to Black-headed Gull but smaller, with thinner black bill and white underside to all primaries. Adult breeding has black hood with white eye-ring; in non-breeding plumage has white head with dark patch on ear-coverts. Juvenile has dark-brown band across wing-coverts, dark trailing-edge to primaries and secondaries and black subterminal tail-band.
Similar species Black-headed Gull is larger, has red bill and dark underside to central primaries.

Tadao Shimba

First-summer. May, Ontario, Canada

Range Counterpart of Black-headed Gull in North America. Breeds mainly in Canada and moves south in winter.
Status in Japan Vagrant. First recorded in 1985 in Ibaragi and there have been a few other sightings from Honshu and Hokkaido.

Saunders's Gull

Larus saundersi 33cm WS87–91cm

Tadao Shimba

Adult breeding. March, Mie

Ran Schols

Adult breeding. May, Beidaihe, China

Description Small gull with pale-grey upperwings, a short, stout, hooked black bill and red legs. Outer four primaries are white with black tips and remaining primaries are white above and black below. In breeding plumage has black hood with white eye-ring; in non-breeding plumage has greyish-black crown and dark patch on ear-coverts. First–winter has brown bar across wing-coverts and a narrow black subterminal tail-band.

Similar species Black-headed Gull is larger and has a longer reddish bill.

Range Breeds in north-eastern China and Mongolia and winters south to southern China.

Status in Japan Common winter visitor to coastal mudflats in northern Kyushu, rare elsewhere.

Tadao Shimba

First-winter. February, Mie

Relict Gull

Larus relictus 46cm WS?

Description Medium-sized gull with sloping head and pale-grey upperwings, black wing–tips with white spot; dark red bill and dark legs. In breeding plumage, acquires black hood and white eye-ring; in non-breeding plumage head is white with grey patch on ear-coverts. First–winter has browner upperwings with dark primaries and black subterminal tail-band. **Similar species** Saunders's Gull is much smaller.

Range Breeds in lakes from north-western Kazakhstan to Mongolia and south-eastern Lake Baikal region. Winters sparsely in southern China and south-east Asia. Rare but regular winter visitor to Korea.
Status in Japan Vagrant. Two sightings on Honshu.

Alister Benn

First-winter. September, Hebei, China

Laughing Gull

Larus atricilla 40cm WS103cm

Description Medium-small gull with dark-grey upperwings with black outer primaries, a long drooping bill, dark legs and white eye-ring. In breeding plumage, adult has black hood and red bill; in non-breeding plumage head is white with dark patch on ear-coverts and dark bill. Adult plumage is attained after three years. First–winter has greyish-brown wash on head and breast, greyish-brown upperwings with dark primaries and secondaries and black tail; second–winter similar to non-breeding adult, but normally with less white on primary tips. **Similar species** Franklin's Gull is slightly smaller, has shorter bill and wings, more white on wing tips and a more pronounced eye-ring.

Norio Yamagata

Second–summer. September, Aichi

Range Breeds in North America from Atlantic coast south to Caribbean. Winters south to South America.
Status in Japan Vagrant. First recorded in 2000 on Iwo-jima with several sightings since then.

Little Gull

Larus minutus 28cm WS64cm

Tomi Muukkonen

Adult breeding. May, Kuusamo, Finland

Aurélian Audevard

First-summer. June, France

Description Adult is a small gull with black bill, red legs, uniform pale-grey upperwings and conspicuous black underwings. Adult breeding has black hood; in non-breeding plumage has white head with grey patch on crown and ear-coverts. First–winter has distinctive black M across upperwings with black subterminal tail-band and white underwings. Buoyant tern-like flight.
Similar species First–winter Ross's Gull is larger and has wedge-shaped tail. First–winter Black-legged Kittiwake is larger, has longer wings and black patch on nape.

Range Breeds from the Baltic east to eastern Transbaikalia; also in Great Lakes regions of North America. Moves south in winter.
Status in Japan Very rare visitor to rivers, lakes and estuaries. First recorded in August 1980 in Hokkaido. Mostly seen in summer.

Sabine's Gull

Larus sabini 34cm WS 90–100cm

Augusto Faustino

Adult breeding. July, Chukotka, Russia

Description Small gull strikingly patterned black, grey and white above with a forked tail. Adult breeding has dark grey hood with black neck ring; white head with dark hindneck in non-breeding plumage. First–winter has greyish-brown hindneck and wing-coverts, and black subterminal tail-band.
Similar species First–winter Black-legged Kittiwake is larger, and has broad black M-shaped pattern across upperwings.

Range Circumpolar. Breeds from the Taimyr Peninsula east along arctic coast to Chukotka; also in polar Nearctic. Winters off western coasts of southern Africa and South America.
Status in Japan Rare passage migrant to pelagic waters off Pacific coast of northern Japan.

Franklin's Gull

Larus pipixcan 36cm WS90cm

Description Adult is a medium-small gull with dark-grey upperwings, red bill and legs and white eye-ring. Black outer primaries with white tips are separated from grey upperwings by white band. In breeding plumage has black hood; in non-breeding plumage has white head with dark patch on ear-coverts and crown and black bill and legs. Adult plumage is attained after two or three years. First–winter has greyish-brown upperwings, a grey saddle, and black subterminal tail-band; second–winter similar to non-breeding adult, but has more black on primaries and lacks the white band.
Similar species Laughing Gull is larger, has longer, drooping bill, longer wings, and a less prominent eye-ring.

Tokio Sugiyama

Second–winter. January, Kyoto

Range Breeds in prairie regions of North America and winters south to South America.
Status in Japan Vagrant. First recorded in 1984 in Kyoto. There are a few other records from Honshu.

Ivory Gull

Pagophila eburnea 44cm WS110cm

Description Adult is a medium-sized, pure-white gull with black legs and yellow-tipped greyish-green bill. First–winter has dark face, black spots on upperparts and black tips to flight feathers and rectrices.
Similar species Other whitish gulls are larger.

Range Breeds in high arctic and mainly sedentary.
Status in Japan Vagrant. A few winter records from central Honshu northwards.

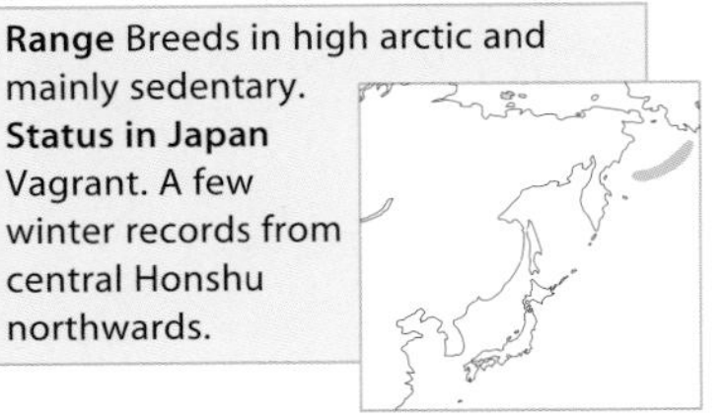

Shigeo Ozawa

First-winter. February, Aomori

Ross's Gull

Rhodostethia rosea 30cm WS84cm

Kazuyasu Kisaichi

Adult breeding. June, Churchill, Canada

Description Small gull with small black bill and distinctive wedge-shaped tail. Adult breeding plumage has narrow black collar and pink-tinged underparts. Adult in non-breeding plumage has white head with dark patch behind eye and grey-tinged neck. First–winter has dark M-shaped mark across upperwings and a black tail-tip. **Similar species** Immature Little Gull is smaller and has square-shaped tail.

Shigeo Ozawa

First-winter. December, Chiba

Range Breeds in high arctic and normally remains in arctic seas in winter.
Status in Japan Rare winter visitor to coastal waters from central Honshu northwards. First recorded in 1974 in Hokkaido with most subsequent records from eastern Hokkaido.

Norio Yamagata

Adult non-breeding. January, Hokkaido

Red-legged Kittiwake

Rissa brevirostris 38cm WS85cm

Description Adult is a medium-sized gull with dark-grey upperwings, triangle-shaped black wing tips, small yellow bill and short red legs. Adult breeding has white head; head grey with dark patch in non-breeding plumage. First–winter has black outer primaries, white inner primaries and outer secondaries, dark patch on ear coverts, dark nape patch, black bill and white tail.
Similar species Adult Black-legged Kittiwake is slightly larger, has paler upperwings, longer bill and black legs; first–winter Black-legged Kittiwake has broad M-shaped pattern on upperwings and black terminal tail-band.

Range Breeds on islands in Bering Sea and remains in the same area in winter.
Status in Japan Vagrant. A few records from Pacific coast of northern Honshu and Hokkaido.

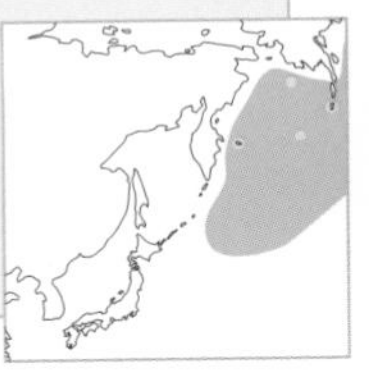

Kazuyasu Kisaichi

Adult breeding. June, St Paul, Alaska

Shigeo Ozawa

First-winter. March, Chiba

Kazuyasu Kisaichi

First-winter. October, Hokkaido

Black-legged Kittiwake

Rissa tridactyla 39cm WS91cm

Tokio Sugiyama

Adult non-breeding. February, Chiba

Description Adult is a medium-sized gull, with grey upperwings, triangle-shaped black wing tips, yellow bill, short black legs and shallow forked tail. Adult breeding has white head; head greyish with a black ear-covert spot in non-breeding plumage. First–winter has broad black M-shaped pattern across upperwings and black tip to tail, black bill and black patch on nape. **Similar species** Common Gull has longer yellow legs and broader black tips on the primaries with white spots. See also Red-legged Kittiwake.

Yoshihito Goto

Adult non-breeding. February, Chiba

Range Breeds on cliffs on coasts of northern Pacific and Atlantic Oceans. Moves south in winter.
Status in Japan Common winter visitor to coastal waters from central Honshu northwards. Often observed in summer in eastern Hokkaido.

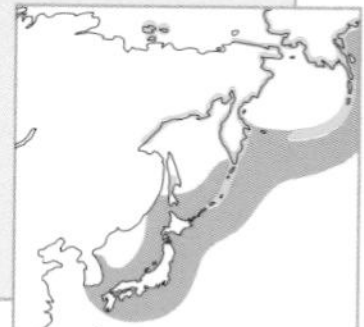

Gull-billed Tern

Gelochelidon nilotica 37.5cm

Himaru Iozawa

Adult breeding. April, Yonaguni-jima

Description Large stocky tern, with pale-grey upperwings, a short, stout black bill, shallow forked tail and longish black legs. In breeding plumage acquires black cap, head white with dark spot on ear-coverts in non-breeding plumage. **Similar species** Common Tern has more deeply forked tail and longer and slender bill.

Range Breeds in temperate Eurasia from Europe east to Transbaikalia; also in Africa and Americas. Winters south in tropical regions.
Status in Japan Rare passage migrant to coastal waters and mudflats. Most often observed in western Japan. Often catches small crabs on coastal mudflats.

Roseate Tern

Sterna dougallii 31cm

Description Medium-sized tern, with pale-grey upperwings and a very long, deeply forked tail. In breeding plumage, black cap, black-tipped red bill and red legs. In non-breeding plumage, white forehead, black bill and brownish-orange legs.
Similar species Common Tern is greyer and has black legs and a shorter tail.

Manabu Tostsuka

Adult breeding. July, Miyako-jima

Range Widely distributed in temperate and tropical seas. Northern populations are migratory. In eastern Asia, breeds on offshore islands from Amami Islands southwards.
Status in Japan Locally common summer visitor to Ryukyu Islands, accidental elsewhere. Breeds in colonies on rocky isles.

Himaru Iozawa

Adult breeding. August, Okinawa-Hontou

Manabu Tostsuka

Adult breeding. July, Miyako-jima

Great Crested Tern

Sterna bergii 45cm

Tokio Sugiyama

Adult. September, Aichi

Other name Swift Tern
Description Large tern, with grey upperwings, large yellow bill, black legs, white forehead and short black crest. Crown black in breeding plumage, becoming white with dark streaks in non-breeding plumage.
Voice Harsh *kreek.*
Similar species Caspian Tern is larger, has red bill. Lesser Crested Tern *S. bengalensis,* (recorded once in Shizuoka) is smaller and has an orange-coloured bill.

Tokio Sugiyama

Juvenile. September, Aichi

Range Breeds on coasts of Indian and western Pacific Oceans.
Status in Japan Uncommon summer visitor to southern Japan. Some breed on Ogasawara and Ryukyu Islands. Post-breeding northbound dispersal is often observed and some regularly occur at a few places in central Honshu.

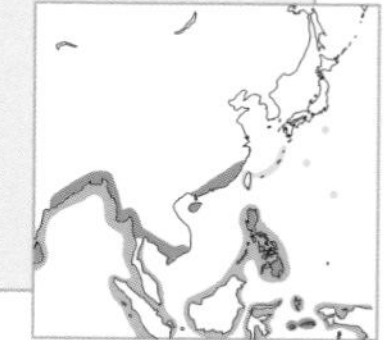

Caspian Tern

Sterna caspia 52.5cm

Teruaki Ishii

First-winter. November, Wakayama

Description Largest tern in the region. Massive red bill with a dark tip, wings grey above and white below with black primaries, black legs and shallow forked tail. Black cap in breeding plumage becomes white with brown streaks in non-breeding plumage.
Similar species Both Great and Lesser Crested Terns are smaller, have different bill colour and lack extensive black on underside of primaries.

Range Breeds in temperate Eurasia from the Balkans east through Black and Caspian Seas to eastern China; also in North America and Australia. Winters in tropics and subtropics.
Status in Japan Very rare visitor to coastal areas from Honshu southwards. Seen almost annually on Okinawa-Hontou.

Black-naped Tern

Sterna sumatrana 30cm

Description Small, very pale tern with long black bill and distinctive black line from eye to nape. In flight shows black leading-edge to primaries and long, deeply forked tail. Juvenile has brown-streaked crown and dark fringes to upperpart feathers.
Similar species Juvenile similar to juvenile Little Tern but larger, and has longer bill and whiter primaries.

Tadao Shimba

Adult. July, Miyako-jima

Range Tropical and subtropical Indian and western Pacific Oceans. Northern population is migratory.
Status in Japan Locally common summer visitor to Ryukyu Islands. Breeds on rocky isles and often seen resting on sandbars with other terns.

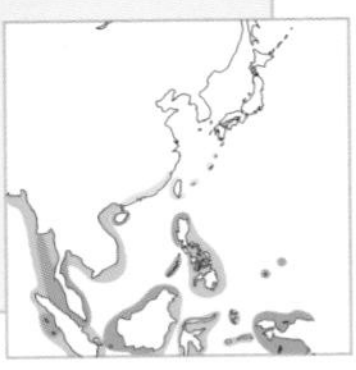

Manabu Totsuka

Adult. July, Miyako-jima

Tadao Shimba

Adults. December, New Caledonia

Common Tern

Sterna hirundo 35.5cm

Tokio Sugiyama

Adult breeding *minussensis*. July, Aichi

Mitsuo Imai

Adult breeding *longipennis*. July, Ibaragi

Description Medium-sized tern, with grey upperwings and deeply forked white tail. In breeding plumage has black cap, greyish-white breast and belly, black bill and red or black legs. In non-breeding plumage has white forehead and underparts and dark carpal bar. Juvenile is similar to non-breeding adult but carpal bar browner. In flight, only inner primaries are translucent. Race *minussensis* from central Siberia has black-tipped red bill and red legs.
Voice Sharp and clipped *kik-kik*.
Similar species See Arctic Tern and Roseate Tern.

Range Widespread throughout northern Eurasia and northern North America. Winters south to warmer areas. In north-east Asia, breeds in eastern Siberia, south to Ussuriland and Kurile Islands.
Status in Japan Common passage migrant in spring and autumn around coastal areas and estuaries. Race *minussensis* is a rare passage migrant.

Mitsuo Imai

Adult non-breeding *minussensis*. July, Ibaragi

Arctic Tern

Sterna paradisaea 33–38cm

Description Medium-sized tern with a short bill and short legs, grey upperwings and deeply forked tail. In breeding plumage has black cap, red bill and greyish underparts. In non-breeding plumage forehead and crown are white and bill black. At rest tail extends beyond the tip of primaries. In flight shows black leading-edge to outer primaries and clear narrow black trailing edge to the white translucent primaries.
Similar species Common Tern has a slightly longer bill, longer legs and shorter tail; only inner primaries are translucent.

Range Circumpolar. Breeds in northern Eurasia and in northern North America; winters in open seas south to Antarctic. Known for long distance migration.
Status in Japan Very rare passage migrant to Pacific coast of Honshu. Occurs in coastal areas and estuaries.

Yasuko Tashiro

Adult breeding. June, Shizuoka

Mitsuo Imai

First-summer. July, Ibaragi

Kazuyasu Kisaichi

First-summer. July, Ibaragi

Little Tern

Sterna albifrons 28cm

Tadao Shimba

Adult breeding. May, Aichi

Description Smallest tern in the region. In breeding plumage has dark-tipped yellow bill, white forehead, black cap and black eye-stripe to base of bill and yellow legs. In non-breeding plumage has black bill and legs, white forehead extends onto crown, eye-stripe does not reach the base of bill. Juvenile similar to non-breeding adult but browner and barred above.

Voice Sharp *krik krik.*

Similar species Aleutian Tern has similar head pattern, but is larger and has black bill and legs.

Tokio Sugiyama

Adult breeding. June, Aichi

Range Widespread in Eurasia and more local in Africa and Australia. Northern populations move south in winter.

Status in Japan Common summer visitor from Honshu southwards. Breeds in coastal areas and at inland waters. Declined since 1980s due to loss of breeding habitat.

Tadao Shimba

Adult breeding. May, Aichi

Aleutian Tern

Sterna aleutica 33cm

Description Medium-sized tern, with grey upperwings and greyish-white underparts. White forehead, black cap and eye-stripe to base of bill, deeply forked tail and black bill and legs. Juvenile has brown head and barred back.
Similar species Non-breeding Common Tern is paler and lacks black eye-stripe. Breeding Little Tern is smaller, paler and has black-tipped yellow bill.

Mitsuo Imai

Adult breeding. June, Ibaragi

Range Breeds on coastal areas in northern Pacific Ocean from far-eastern Russia south to Sakhalin; also in Alaska. Winters south to warmer oceans.
Status in Japan Very rare visitor to coastal waters in northern Japan. Mainly recorded in spring and summer.

Grey-backed Tern

Sterna lunata 33cm

Description Medium-sized tern, with grey upperwings, black bill and legs. Black cap with white forehead extending over eye.
Similar species Bridled Tern has dark brownish-grey upperparts and longer tail.

Range Breeds in central Pacific Ocean from Fiji north to Hawaiian Islands.
Status in Japan Vagrant. A few records from Minami-Torishima and Iwo-jima.

Manabu Totsuka

Adult. April, Hawaii, USA

Sooty Tern

Sterna fuscata 40.5cm

Tadao Shimba

Adult. January, Lord Howe, Australia

Description Medium-sized tern, black above with white underparts, a white forehead, black eye-stripe and black bill and legs. Juvenile is dark-brown overall with white barring on upperparts.
Similar species Bridled Tern is slightly smaller, browner and its white supercilium extends behind eye.

Range Widely distributed in tropical and subtropical oceans.
Status in Japan Locally common summer visitor to Minami-Torishima, Ogasawara Islands and Nakanougan-jima; accidental elsewhere.

Bridled Tern

Sterna anaethetus 35.5cm

Tokio Sugiyama

Adult. July, Miyako-jima

Description Medium-sized tern, dark grey-brown above with white underparts. White forehead, black eye-stripe and black bill and legs.
Similar species Sooty Tern has blacker upperparts and white supercilium does not extend behind eye.

Range Widely distributed in tropical and subtropical oceans.
Status in Japan Locally common summer visitor to southern Ryukyu Islands. Breeds in colonies on rocky isles and coral reefs.

White-winged Tern

Chlidonias leucopterus 23.5cm

Description Small tern with shallow forked tail. In breeding plumage has greyish-white wings and white tail contrast with black head, breast and back, also black underwing-coverts. In non-breeding plumage has mainly white head with black patch at rear of eye. Juvenile is similar to non-breeding adult but mantle browner. Flight is buoyant and restless. Flies low over water and feeds by dipping to the water surface, rarely dives.

Similar species Non-breeding Black Tern has longer, more pointed bill and has dark breast patches. Non-breeding Whiskered Tern has stouter bill and streaked crown.

Tadao Shimba

Adult breeding. April, Melbourne, Australia

Akira Yamamoto

Juvenile. September, Okinawa-Hontou

Range Palearctic region; breeds locally from south-eastern Europe east to north-eastern China and Ussuriland. Winters south to Africa, south-east Asia and Australia. Rare passage migrant in Korea.

Status in Japan Uncommon passage migrant to coastal waters and freshwater marshes.

Tadao Shimba

First-winter. April, Melbourne, Australia

Whiskered Tern

Chlidonias hybridus 25cm

Mitsuo Imai

Adult breeding. July, Ibaragi

Shuichi Tsuno

Juvenile. September, Wakayama

Description In breeding adult, white cheeks contrast with black cap and breast. Greyish wings and tail, red legs and dark-tipped red bill. In non-breeding plumage has white forehead, streaked crown and white underparts. Juvenile has dark-brown crown and nape; mantle and scapulars black and gingery-brown. Flies low over water and feeds by dipping to the water surface; rarely dives.

Similar species Non-breeding White-winged Tern has black patch at rear of eye and slimmer bill. Non-breeding plumage Black Tern has darker upperwings and dark patches on sides of breast.

Range Breeds in temperate Eurasia from Europe east to north-eastern China, also in Africa and Australia. Winters mainly in tropical regions. Rare passage migrant to Korea.

Status in Japan Uncommon passage migrant in coastal waters and freshwater marshes. Some winter in southern Japan.

Akira Yamamoto

Adult breeding. June, Aichi

Black Tern

Chlidonias niger 24cm

Description Small tern with black bill and legs. In breeding plumage has black head and breast and dark-grey upperwings and tail.
In non-breeding plumage has white face with dark crown and dark patch at rear of head and dark patches on the side of breast. Juvenile has dark-greyish mantle and carpal bars.
Similar species In non-breeding plumage separated from both White-winged Tern and Whiskered Tern by longer pointed bill and dark patches on the side of breast.

Tomi Muukkonen
Adult breeding. June, Lubans, Latvia

Range Breeds from western Europe east to western China and southern Russia; also in North America. Winters mainly in Africa and South America.
Status in Japan Very rare passage migrant in coastal waters and freshwater marshes.

Kazuyasu Kisaichi
First-summer. August, Ibaragi

Black Noddy

Anous minutus 36cm

Description Medium-sized dark tern, with silvery-white forehead; long, thin bill, wedge-shaped tail and black underwings.
Similar species Brown Noddy is larger and browner, has shorter bill, also forehead greyer and underwings browner.

Kazuyasu Kisaichi
Adults. July, Miyako-jima

Range Breeds in trees on islands in tropical and subtropical Pacific and Atlantic Oceans.
Status in Japan Rare summer visitor to Ryukyu Islands, Iwo-jima and Minami-Torishima. There are breeding records from Iwo-jima.

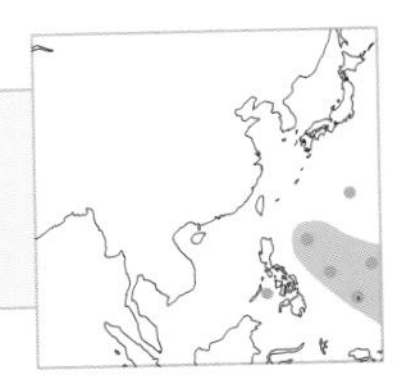

Brown Noddy

Anous stolidus 39cm

Mnabu Totsuka

Adult *pullus* and chick. July, Miyako-jima

Description Medium-sized, sooty-brown tern with greyish-white forehead, broad wedge-shaped tail and brown underwings. Takes small fish by dipping to the water surface and rarely dives like other terns. **Similar species** Black Noddy is smaller, darker and has thinner, longer bill; also a darker underwing.

Kazuyasu Kisaichi

Adult *pullus*. July, Miyako-jima

Range Nearly cosmopolitan in tropical and subtropical oceans. **Status in Japan** Locally common summer visitor to Ogasawara Islands, Iwo-jima and Minami-Torishima. Breeds in colonies on rocky isles and coral reefs. Race *pullus* breeds on Yaeyama Islands; the more widely distributed *pileatus* breeds on the Ogasawara Islands.

Manabu Totsuka

Adult *pullus*. July, Miyako-jima

White Tern

Gygis alba 27.5cm

Other names Fairy Tern, Angel Tern
Description Unmistakable all-white tern; has dark brown eye with conspicuous black eye-ring, slightly upcurved black bill with blue at base and black feet with blue webs. Juvenile has dark patch behind eye and greyish markings on nape and mantle. Lays single egg on branch of tree, often very approachable.
Similar species Black-naped Tern can appear all white at a distance.

Tadao Shimba

Adult. January, Lord Howe Island, Australia

Range Widely distributed in tropical and subtropical Pacific, Indian and Atlantic Oceans.
Status in Japan Vagrant. Most records are birds brought in by typhoons.

Little Auk

Alle alle 20cm

Description Small, plump auk with stubby bill and short neck. In breeding plumage, black head, breast and back contrast with white underparts. In non-breeding plumage has white throat and sides of neck. In flight shows white trailing edges to secondaries and very fast wingbeats.
Similar species Ancient Murrelet has thinner yellowish bill and greyish-black back.

Himaru Iozawa

Adult breeding. May, Shizuoka

Range Breeds in the Arctic and winters in northern Atlantic.
Status in Japan First recorded in 1992 on Okinawa-Hontou, with a few other sightings since then.

Common Guillemot

Uria aalge 44cm

Nobuo Shiraki

Adult non-breeding. December, Hokkaido

Description Medium-sized auk with dark sooty-brown head, neck and back contrasting with white underparts. In non-breeding plumage, face and cheeks become white with a black line extending from eye across cheeks. In flight shows white trailing edges to secondaries.
Voice Only heard at breeding colonies; high growling *arr* or *ooarr.*
Similar species Brünnich's Guillemot is blacker, has a thicker bill with white line on upper mandible and is darker-headed in non-breeding plumage.

Range Breeds on coasts and islands of northern Pacific and Atlantic oceans. Winters at sea south to the temperate regions.
Status in Japan Common winter visitor to offshore waters from central Honshu northwards. Occasionally visits coastal areas and fishing ports in northern Japan. A large colony existed on Teuri-tou off the west coast of Hokkaido until 1980s but now only a few remain.

Brünnich's Guillemot

Uria lomvia 44cm

Kazuyasu Kisaichi

Adult non-breeding. December, Hokkaido

Other name Thick-billed Murre
Description Similar to Common Guillemot, but upperparts blacker and bill thicker with a white line on cutting edge of upper mandible. In non-breeding plumage, greyish-white cheeks and throat give dark-headed appearance. In flight shows white trailing edges to secondaries.
Similar species Common Guillemot has browner back, lacks white line on bill and has a prominent black line across white cheek in non-breeding plumage.

Range Breeds on coasts of Bering Sea south to Kamchatka Peninsula, Sakhalin and Kuril Islands; also in northern Siberia, arctic Canada, Greenland and Iceland. Winters in northern waters.
Status in Japan Uncommon winter visitor to offshore waters in northern Japan. Occasionally visits fishing ports and coastal areas in Hokkaido.

Spectacled Guillemot

Cepphus carbo 37cm

Description Pigeon-sized auk with long black bill and red legs. In breeding plumage is sooty-black overall with distinctive white spectacles and white spot at the base of bill. In non-breeding plumage has white eye-ring and white throat, foreneck and underparts. In flight shows entirely sooty-black wings.
Voice High, persistent whistles *peeeeh peeeh*.
Similar species Non-breeding Pigeon Guillemot has mottled dark-brown and white head, hindneck and back.

Norio Yamagata

Adult breeding. June, Aomori

Range Breeds on coasts of Okhotsk Sea, Kamchatka Peninsula, south to Sakhalin, Kurile Islands and Korea. In winter, remains in the same area on open seas.
Status in Japan Locally common resident in northern Japan and breeds on rocky shores. Winters from central Honshu northwards. Numbers have declined in recent years.

Kazuyasu Kisaichi

Adult non-breeding. March, Hokkaido

Kazuyasu Kisaichi

Adult breeding. July, Hokkaido

Pigeon Guillemot

Cepphus columba 33cm

Kazuyasu Kisaichi

Adult breeding. June, Alaska

Description Pigeon-sized auk with slender neck and black bill. In breeding plumage sooty-black, with distinctive large white upperwing-patch bisected by a dark wedge. In non-breeding plumage has white face with dark eye-patch, white–mottled dark-brown hindneck and back, white foreneck and underparts. In flight shows dark bar across white wing-coverts.
Similar species Spectacled Guillemot lacks white patches on wing-coverts.

Range Breeds on coasts and islands of the northern Pacific from Kamchatka east to western Alaska, south to Kurile Islands. Winters south to Hokkaido and California.
Status in Japan Rare winter visitor to offshore waters in northern Japan.

Kittlitz's Murrelet

Brachyramphus brevirostris 24cm

Gary Luhm

Adult breeding. Alaska.

Description Small auklet with a short, dark bill. In breeding plumage has greyish-brown back mottled and barred with white, and whitish underparts heavily marked brown. In non-breeding plumage has sooty-grey cap, white face and underparts; sooty-grey hindneck and back with white scapulars. White outer tail feathers prominent in flight.
Similar species Long-billed Murrelet has larger bill and darker face; also lacks white on outer tail feathers.

Range Northern Pacific. Breeds on islands of the Bering Sea from Chukotski Peninsula east to southern Alaska. Winters on open seas south to Kuriles.
Status in Japan A few unconfirmed sightings in northern Japan.

Long-billed Murrelet

Brachyramphus perdix 24.5cm

Description Small auk with long black bill and yellowish-brown legs. In breeding plumage, dark-brown with whitish cheeks and pale throat. In non-breeding plumage has greyish-brown head, hindneck and back with white scapulars, prominent white eye-ring and white underparts.
Similar species Non-breeding Ancient Murrelet has shorter yellow bill and darker chin.

Tokio Sugiyama

Non-breeding. December, Sizuoka

Range Breeds on coasts of the Okhotsk Sea, Kamchatka south to Sakhalin and the Kurile Islands. Winters south to Korea and northern Japan.
Status in Japan Uncommon winter visitor to offshore waters in northern Japan. Breeds in trees in dense forests and hard to detect. One confirmed breeding record in eastern Hokkaido and may still breed there.

Tokio Sugiyama

Non-breeding. December, Shizuoka

Kazuyasu Kisaichi

Non-breeding. December, Sizuoka

Japanese Murrelet

Synthliboramphus wumizusume 24cm

Teruaki Ishii

Adult breeding. February, Miyazaki

Description Similar to Ancient Murrelet, but has black crest on crown and distinctive broad white stripes on sides of head, black face and hindneck, dark-grey back and white underparts. In non-breeding plumage lacks crest and face whiter. **Similar species** Ancient Murrelet has pale thicker bill and lacks white stripe on head in breeding plumage; face darker in non-breeding plumage.

Nobuo Shiraki

Adult non-breeding. January, Ehime

Range Breeds on small islands in Japan and southern Korea. Winters on open seas.
Status in Japan Uncommon resident on offshore waters from central Honshu southward. Breeds locally on rocky coasts and small islands from Izu Islands south to southern Kyushu. Winters on open waters around breeding areas.

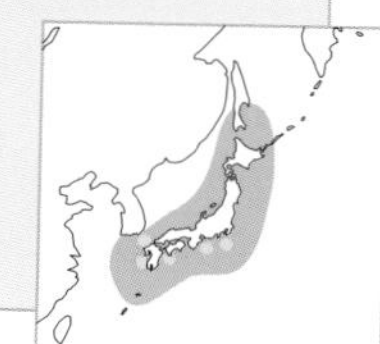

Nobuo Shiraki

Adult non-breeding. January, Ehime

Ancient Murrelet

Synthliboramphus antiquus 25cm

Description Small auk with short yellow bill, bluish-grey legs, black head and greyish-black back. In breeding plumage has black throat and white plumes fringing the crown. In non-breeding plumage has greyish throat and white extending to behind ear-coverts.
Voice Soft whistle *cheerp*.
Similar species Long-billed Murrelet has longer thin black bill and white chin.

Nobuo Shiraki

Adult breeding. January, Hokkaido

Range Northern Pacific. Breeds from Kamchatka Peninsula east through the Aleutian Islands to western Alaska, south to Sakhalin, Kurile Islands, Korea and Japan.
Status in Japan Most common murrelet in Japan. Breeds on rocky coasts in northern Japan and winters to southern Japan. In winter, small flocks are often seen from the coast.

Tokio Sugiyama

Adult non-breeding. December, Sizuoka

Tokio Sugiyama

Adult non-breeding. December, Hokkaido

Least Auklet

Aethia pusilla 15cm

Kazuyasu Kisaichi

Adult breeding. June, St Paul

Description Tiny auklet with small, dark, red-tipped bill. In breeding plumage has black head and back, white eyes with single white plume behind eye, white throat, white underparts heavily mottled with grey and white streaks on scapulars. In non-breeding plumage has white cheeks and underparts.

Similar species Other murrelets and auklets are larger.

Kazuyasu Kisaichi

Adult non-breeding. December, Hokkaido

Range Northern Pacific. Breeds on coasts of Chukotka and western Alaska, south to the Aleutians. Winters on open seas.

Status in Japan Common winter visitor to offshore waters in northern Japan.

Parakeet Auklet

Cyclorrhynchus psittacula 25cm

Kazuyasu Kisaichi

Adult breeding. June, St Paul

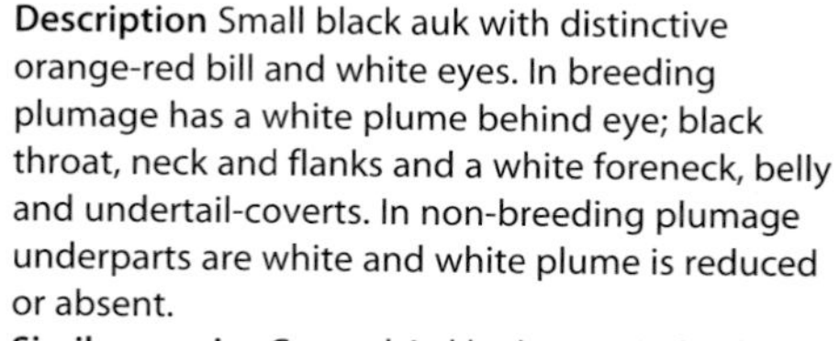

Description Small black auk with distinctive orange-red bill and white eyes. In breeding plumage has a white plume behind eye; black throat, neck and flanks and a white foreneck, belly and undertail-coverts. In non-breeding plumage underparts are white and white plume is reduced or absent.

Similar species Crested Auklet has entirely black body and forward–curling crest on forehead. Whiskered Auklet is smaller with three white plumes on face.

Tadao Shimba

Adult non-breeding. August, Iwate

Range Northern Pacific. Breeds on Bering Sea and Aleutian islands, south to Kamchatka and northern Kurile.

Status in Japan Rare winter visitor to offshore waters in northern Japan.

Crested Auklet

Aethia cristatella 24cm

Description Small black auk with distinctive crest curling forward from forehead. In breeding plumage has narrow white plume behind eye and bright red bill. In non-breeding plumage bill is smaller and duller, and crest shorter.
Similar species Whiskered Auklet is smaller, has three white plumes on face and white undertail-coverts.

Mitsuo Imai

Adult non-breeding. March, Ibaragi

Range Northern Pacific. Breeds on coasts and islands of Bering Sea, south to Sakhalin and Kurile Islands. Winters on open seas.
Status in Japan Common winter visitor to offshore waters in northern Japan. Large flocks often seen from the Pacific coastal ferries.

Whiskered Auklet

Aethia pygmaea 20cm

Description Small, dark-brown auklet with three white plumes on face; red bill with white tip and thin crest curling forward from forehead; pale undertail-coverts. Plumes and crest shorter in non-breeding plumage.
Similar species Crested Auklet is larger, has single plume on face and dark undertail-coverts.

Range Northern Pacific. Breeds on rocky shores and cliffs on the Aleutians and on islands off Siberian coast south to the Kuriles. Mainly sedentary, but winters on open seas.
Status in Japan Rare winter visitor to offshore waters in northern Japan.

Martin Hale

Adult breeding. Kamchatka.

Rhinoceros Auklet

Cerorhinca monocerata 38cm

Kazuyasu Kisaichi

Adult breeding. July, Hokkaido

Description Large dark-brown auklet with large head, short neck and pale yellow legs. In breeding plumage has white plumes on cheeks and behind eye and orange bill with pale yellowish-brown horn-shaped comb at the base. Grey breast, pale belly and undertail-coverts. In non-breeding plumage bill is duller and hornless, and there is reduced white on face.
Similar species Both Tufted Puffin and Horned Puffin have much heavier bills.

Kazuyasu Kisaichi

Adult breeding. July, Hokkaido

Range Northern Pacific. Breeds from Sakhalin and Kurile Islands south to northern Japan; also on Aleutian Islands and western Alaska south to northern California. Winters on open seas.
Status in Japan Common resident in offshore waters in northern Japan. Breeds on islands off Hokkaido and northern Honshu. Abundant on Teuri-tou off west coast of Hokkaido in breeding season.

Norio Yamagata

Adult breeding. June, Hokkaido

Horned Puffin

Fratercula corniculata 38cm

Description Unmistakable black auk with white underparts and red legs. In breeding plumage has white face with small black horn above eye and black eye-line behind eye. Massive yellow bill with red tip. In non-breeding plumage has grey face and grey bill with dull orange tip.
Similar species Tufted Puffin has a different facial pattern and dark underparts.

Kazuyasu Kisaichi

Adult breeding. June, St Paul, Alaska

Range Northern Pacific. Breeds on coasts of the Okhotsk Sea and Kamchatka east through islands of Bering Sea to western Alaska, south to Sakhalin and northern Kurile Islands. Winters on open seas.
Status in Japan Rare winter visitor to offshore waters in northern Japan. A few summer records in Hokkaido.

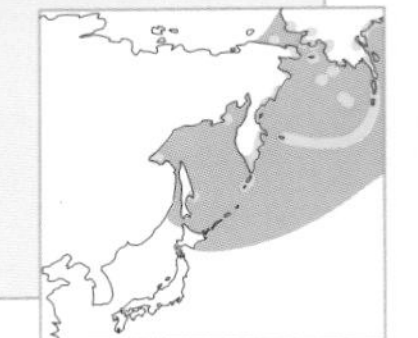

Tufted Puffin

Fratercula cirrhata 38cm

Description Stocky black auk with characteristic bill and red legs. In breeding plumage has white face with distinctive long yellow plumes behind eye, massive reddish-orange bill with yellow at base. In non-breeding plumage has grey face and dull orange bill.
Similar species Horned Puffin has different facial pattern and white underparts.

Haruki Moriyama

Adult breeding. July, Hokkaido

Range Northern Pacific. Breeds on coasts of the Okhotsk Sea and Kamchatka east through islands of Bering Sea to western Alaska and British Columbia, south to Sakhalin and eastern Hokkaido. Winters on open seas.
Status in Japan Uncommon winter visitor to offshore waters in northern Japan. Bred regularly in eastern Hokkaido until the 1980s, then seriously declined; only a few breeding pairs remain.

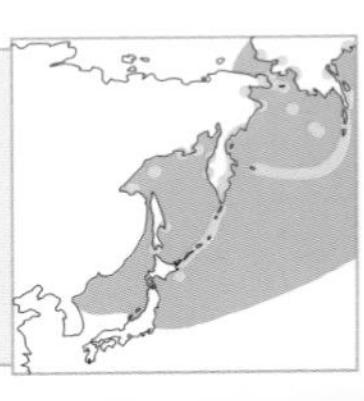

Pallas's Sandgrouse

Syrrhaptes paradoxus 37cm

Nobuo Shiraki

Adults. June, Gobi Desert, Mongolia

Description Plump yellowish-brown bird, with long, needle-like central tail feathers. Distinctive black patch on belly and white undertail-coverts. Male has orange head and throat and grey breast, female has yellowish head with fine dark streaks on crown and ear-coverts.

Similar species Only sandgrouse in north-eastern Asia.

Nobuo Shiraki

Adult ♂. Gobi Desert, Mongolia

Range Breeds from central Asia east to Mongolia and north-eastern China. Vagrant to Korea. Inhabits deserts and grassy steppes.

Status in Japan Very rare visitor; no recent records.

Hill Pigeon

Columba rupestris 34cm

Hyun-tae Kim

Adult. August, Goorye, South Korea

Description Similar to Domestic Pigeon, but has distinctive white tail-band and iridescent neck-patches. Bluish-grey upperparts with two black wing-bars, dark bill with pinkish cere, purplish-brown breast and reddish legs.

Range Central Asia east to Ussuriland, northern China and Korea. Occurs on cultivated fields, mountain cliffs and around human habitation.

Status in Japan No records.

Japanese Wood Pigeon

Columba janthina 40cm

Description Large sooty-black pigeon with relatively small head, purplish crown and hindneck, pale greenish-blue bill with yellow tip and red legs. Sexes alike.
Voice Deep, hoarse *cooooooo, cooooooo.*

Range Resident in southern Japan and on islands off southern Korea.
Status in Japan Uncommon resident in broad-leaved evergreen woodlands, from central Honshu south to Ryukyu and Ogasawara Islands. Inhabits mainly coastal areas and forested offshore islands. Race *nitens* on the Ogasawara Islands is critically endangered.

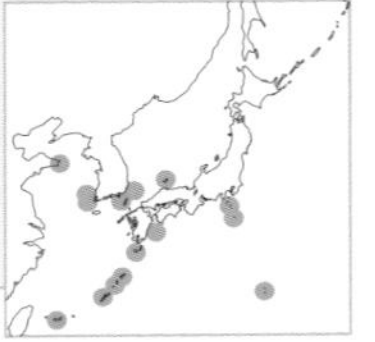

Tadao Shimba

Adult *stejnegeri*. July, Miyako-jima

Akira Yamamoto

Adult *stejnegeri*. May, Miyako-jima

Tadao Shimba

Adult *stejnegeri*. July, Miyako-jima

Oriental Turtle Dove

Streptopelia orientalis 32–35cm

Tadao Shimba

Adult *orientalis*. February, Aichi

Description Purplish-grey head, neck and underparts. Bluish-grey and black striped patches on sides of neck. Dark-brown wings with rufous edges to wing-coverts, dark tail with grey-tipped rectrices. Two races in Japan; race *stimpsoni* on Ryukyu Islands is darker overall.
Voice Slow pitched, low and distinctive *door door paw paw*.

Tadao Shimba

Adult *stimpsoni*. April, Ishigaki-jima

Range Eastern Eurasia from the Urals east to the Okhotsk coast and Sakhalin, south to the Indian subcontinent and south-east Asia. Northern populations are migratory.
Status in Japan Common resident in wooded areas, parks and gardens. Summer visitor to Hokkaido.

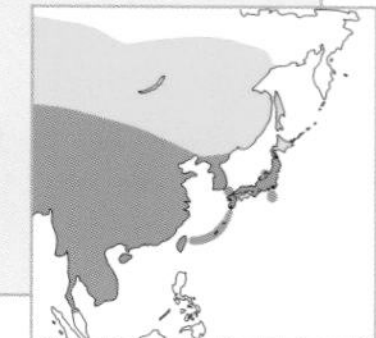

Eurasian Collared Dove

Streptopelia decaocto 31–33cm

Mitsuo Imai

Adult. November, Saitama

Description Slim and long-tailed dove. Pale greyish-brown overall with darker wings. Black patch on sides of neck and white tips to outer tail feathers.
Voice Muted *coo-coo.*
Similar species Female Red Collared Dove is smaller and browner, and has shorter tail.

Range Resident in temperate Eurasia from Europe east through the Indian subcontinent to eastern China and Korea.
Status in Japan Resident in Saitama, Chiba and Tochigi where locally common resident in cultivated fields and urban wooded areas. Presumed introduced from southern Asia in the Edo era.

Red Collared Dove

Streptopelia tranquebarica 23cm

Description Small dove with grey head and black nape band. Reddish-brown back, dark-brown primaries, greyish rump and white outer tail feathers, purplish-brown breast and belly and white lower belly. Female is browner than male.
Similar species Collared Dove is larger and greyer, and has longer tail.

Range Indian subcontinent east to eastern China, Taiwan and northern part of south-east Asia. Northern populations are migratory.
Status in Japan Rare visitor to cultivated fields and grassy open fields. Most often seen on the Ryukyu Islands.

Akira Yamamoto

Adult ♂. March, Aichi

Tokio Sugiyama

Immature ♂. February, Aichi

Emerald Dove

Chalcophaps indica 25cm

Description Small dove with purplish-brown head, back and underparts. Bronzed emerald-green wings with black primaries, dark rump and tail with two grey bars across lower back, reddish bill and purplish legs. Male has grey crown, silvery-white lores and supercilium and white shoulder-patch. Female is duller and has pale-brown lores.
Voice Six or seven low-pitched moaning *coo,* starting softly, and then rising; also a nasal *hoo-hoo-hoo.*

Range Breeds from the Indian subcontinent east to southern China and Taiwan, south to south-eastern Asia and eastern Australia.
Status in Japan Locally common resident on Miyako and Yaeyama Islands. Inhabits evergreen subtropical forests.

Manabu Totsuka

Adult ♂. September, Ishigaki-jima

Kazuyasu Kisaichi

Adult ♀. October, Ishigaki-jima

White-bellied Green Pigeon

Treron sieboldii 33cm

Teruaki Ishii

Adult ♂. August, Shizuoka

Teruaki Ishii

Adult. ♀. August, Shizuoka

Description Yellow-green head and breast, greyish-green back, and reddish-purple wing-coverts with dark primaries. Greyish-blue bill, yellowish-white belly and white undertail-coverts with dark streaks. Female has greenish-brown wing-coverts.

Voice Variable howling and a series of cooing notes.

Similar species Whistling Green Pigeon is darker green and has a less wedge-shaped tail.

Range Resident in eastern Asia from Japan south to south-eastern China, Taiwan and northern Vietnam.

Status in Japan Uncommon resident in evergreen and deciduous forests from Kyushu northward. Summer visitor to Hokkaido and a rare winter visitor to the Ryukyu Islands. This species is known to drink salt water in summer when flocks of hundreds can be seen around rocky coasts.

Tadao Shimba

Adults. August, Kanagawa

Whistling Green Pigeon

Treron formosae 35cm

Description Dark yellowish-green head, back and underparts with heavy dark streaks on undertail-coverts, and bright-blue bill. Sexes are similar, but male has purplish-brown tinge to median and lesser-coverts.
Voice Similar to White-bellied Green Pigeon, but shorter *poh-oh.*
Similar species White-bellied Green Pigeon has whiter belly and lighter greenish head and breast.

Kazuyasu Kisaichi
Adult ♂. July, Miyako-jima

Range Resident in southern Japan south to Taiwan and northern Philippines.
Status in Japan Common resident in evergreen forests and wooded urban areas from Yakushima southwards.

Manabu Totsuka
Adult ♂. April, Iriomote-jima

Kiyoto Ogata
Adult ♀. May, Okinawa-Hontou

Indian Cuckoo

Cuculus micropterus 31–33cm

Nobuo Shiraki

Adult. May, Hegura-jima

Teruaki Ishii

Adult. May, Hegura-jima

Description Dark bluish-grey head and breast, and dark brown back and tail. Dark tail with black subterminal band and white tip, whitish underparts with heavy dark barring; white undertail.
Voice Loud and distinctive four syllables with fourth syllable descending *farh-farh-farh-four.*
Similar species Both Common and Oriental Cuckoo have bluish-grey backs and lack black sub-terminal tail band.

Range Breeds from the Indian subcontinent and Himalayas east to south-east Asia, north to the Amur Basin, northern China, Ussuriland and Korea. Northern populations are migratory.
Status in Japan Very rare spring and summer visitor to western Japan and off-shore islands in Sea of Japan. Breeding has not yet been confirmed.

Nobuo Shiraki

Adult. May, Hegura-jima

Oriental Cuckoo

Cuculus saturatus 32cm

Description Dark bluish-grey head, breast and back with dark-brown wings and tail. Whitish belly boldly barred black, blackish bill with yellowish base and yellow legs. Females have a brown morph which is heavily barred rufous and black above.
Voice Monotonous, muted and resonant *po-po po-po po-po.*
Similar species Common Cuckoo has lighter bluish-grey back and narrower bars on underparts.

Norio Yamagata

Adult ♂. May, Hegura-jima

Range Breeds from western Siberia east to Kamchatka, Sakhalin, Ussuriland and north-eastern China. Winters in south-east Asia and northern Australia.
Status in Japan Common summer visitor to mountain forests from Kyushu northwards. Often observed in city parks on fall migration. Mainly parasitises Eastern Crowned Warbler.

Kazuyasu Kisaichi

Adult ♀. May, Hegura-jima

Yoshio Osawa

Adult ♀ brown morph. May, Aomori

Little Cuckoo

Cuculus poliocephalus 26cm

Norio Yamagata

Adult ♂. May, Hegura-jima

Description Dark bluish-grey head, breast and back with dark-brown wings and tail. Whitish belly with widely spaced dark bars, dark bill with yellow base and yellowish legs. Females have a brown morph which is heavily barred rufous and black above.

Voice Loud and distinctive *hyut hyut tyo-kyo-kyo,* often calls in flight.

Similar species Both Common and Oriental Cuckoos are larger and have more heavily barred underparts.

Kazuyasu Kisaichi

Adult ♂. May, Hegura-jima

Range Breeds from the Himalayas through China to Ussuriland and Korea. Winters in southern Asia.

Status in Japan Common summer visitor to wooded areas from Okinawa north to southern Hokkaido. Mainly parasitises Japanese Bush Warbler and Winter Wren.

Akira Yamamoto

Adult ♂. June, Aichi

Northern Hawk Cuckoo

Cuculus hyperythrus 32cm

Other name Rufous Hawk Cuckoo
Description Hawk-like cuckoo with dark-grey head, back and wings. Has yellow eye-ring, dark bill with yellow base and pale-rusty breast and belly, grey tail with black bands and rusty tip, and yellow legs. Juvenile has dark-brown back and pale underparts with dark streaks.
Voice Loud and distinctive *zheeyou-echee;* often calls in flight.

Range Breeds in eastern Asia from eastern Russia north to the Amur Basin, China from Sichuan east to Lower Yantze, Ussuriland and Korea. Northern populations are migratory and winter in south-eastern Asia.
Status in Japan Uncommon summer visitor to mountain forests from Kyushu northwards. Mainly parasitises Siberian Blue Robin, Japanese Robin and Red-flanked Bluetail.

Shigeo Ozawa

Adult. May, Hegura-jima

Yoshihito Goto

First-summer. May, Hegura-jima

Shigeo Ozawa

Adult. May, Hegura-jima

Common Cuckoo

Cuculus canorus 33–36cm

Tadao Shimba

Adult ♂. May, Hegura-jima

Description Bluish-grey head, breast and back with dark-brown wings and tail. White belly finely barred dark-brown. Dark-brown bill with yellowish base and yellow legs. Females have a brown morph that is heavily barred rufous and black above.
Voice Distinctive, familiar call *cuck-coo.*
Similar species Oriental Cuckoo is darker and has broader bars on belly. Lesser Cuckoo is smaller and has sparsely barred belly.

Range Breeds in temperate Eurasia from Europe east to Kamchatka, Sakhalin and Ussuriland, south to the Himalayas and southern China. Winters in tropical Africa and southern Asia.
Status in Japan Common summer visitor to grassy meadows and forest edges from Kyushu northwards. Mainly parasitises Oriental Reed Warbler, Azure-winged Magpie, Bull-headed Shrike and Meadow Bunting.

Teruaki Ishii

Adult ♀. May, Hegura-jima

Tadao Shimba

Adult ♀. May, Hegura-jima

Collared Scops Owl

Otus bakkamoena 24–25cm

Description Small dark-brown owl with short ear-tufts and orange eyes. Greyish-brown underparts finely streaked and lightly barred dark-brown, tarsus entirely feathered. Two races in Japan; *pryeri* on Ryukyu Islands has buff plumage and unfeathered tarsi.
Voice Low, monotonous hooting *hoo-hoo-hoo-hoo.*
Similar species Oriental Scops Owl is smaller, has yellow eyes, greyish-brown back and unfeathered lower tarsus.

Range Breeds from Ussuriland and Sakhalin south through eastern China and Korea to the Greater Sunda Islands.
Status in Japan Uncommon resident in deciduous and mixed forests. Northern populations move to warmer areas in winter; occasionally found roosting in tree hollows in city parks.

Shigeo Ozawa

Adult *semitorques*. November, Tokyo

Norio Yamagata

Adult *semitorques*. November, Niigata

Kazuyasu Kisaichi

Adult *pryeri*. May, Okinawa-Hontou

Ryukyu Scops Owl

Otus elegans 22cm

Tadao Shimba

Adult. April, Ishigaki-jima

Description Small, reddish-brown owl with small ear-tufts and yellow eyes. Has a reddish-brown head, nape and breast; greyish-brown back and pale underparts finely streaked dark brown.
Voice Male's territorial call is a soft whistle *ohh, ohh, ohh*, repeated steadily and ends with upward inflection; female then replies *ey-eh ey-eh*.
Similar species Oriental Scops Owl is slightly smaller, greyer overall and has broad streaks on underparts. Collared Scops Owl is larger and has orange eyes.

Range Ryukyu Islands, Lanyu Island off south-eastern Taiwan and Batan Island in the Philippines.
Status in Japan Common resident in wooded areas from the Amami Islands southwards, often easily detected by spotlighting on Yaeyama Islands.

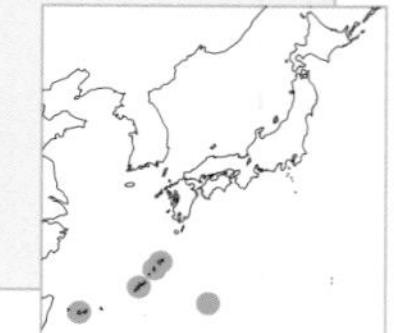

Kazuyasu Kisaichi

Adult. July, Ishigaki-jima

Akira Yamamoto

Adult. June, Miyako-jima

Oriental Scops Owl

Otus sunia 18–21cm

Description Small greyish-brown owl with short ear-tufts and yellow eyes. Pale underparts finely streaked and barred dark-brown. Occurs in two colour morphs, grey-brown and rufous.
Voice Call has two or three syllables: *coh-coh* or *co-co-coh*.
Similar species Collared Scops Owl is larger, darker and has orange eyes.

Norio Yamagata

Adult grey morph. May, Hegura-jima

Range Breeds from Sakhalin and Ussuriland south through eastern China and Korea to northern part of south-eastern Asia and the Indian subcontinent. Northern populations are migratory.
Status in Japan Uncommon summer visitor to mountain forests from Kyushu northwards. Some winter in southern Japan.

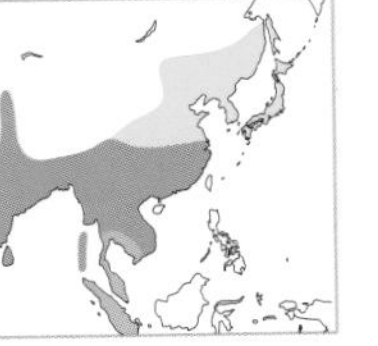

Teruaki Ishii

Adult rufous morph. May, Nara

Northern Hawk Owl

Surnia ulula 38cm

Description Long-tailed, hawk-like owl with yellow eyes and bill, white facial discs with a black border; greyish-brown above with white spots and white scapular patches, white underparts finely barred brown. Active by day and often perches on top of spruce trees or telephone poles looking for prey.

Range Breeds across the boreal zone of Eurasia east to north-eastern Russia and Sakhalin; also in northern North America. Some move south in winter; a rare visitor to northern China and Korea.
Status in Japan No records.

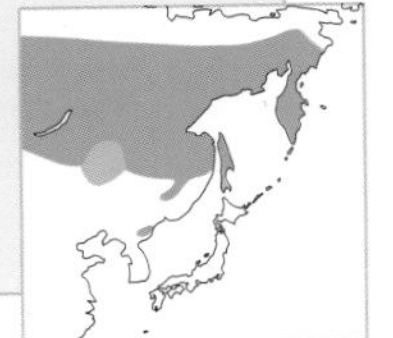

Tadao Shimba

Adult. February, Ontario, Canada

Snowy Owl

Nyctea scandiaca 53–66cm

Tadao Shimba

Adult ♀. January, Ontario, Canada

Description Large, round-headed white owl. Adult male is almost entirely white. Female is larger and more heavily marked with dark spots and bars; juvenile is heavily spotted and barred grey.

Range Circumpolar. Breeds in Arctic Eurasia and North America. Mostly remains in breeding range in winter. Erratic irruptions south occur when food supplies are short.
Status in Japan Rare winter visitor to Hokkaido. Observed in large open fields and coastal areas. A few summer records from the Daisetsu Mountains in central Hokkaido.

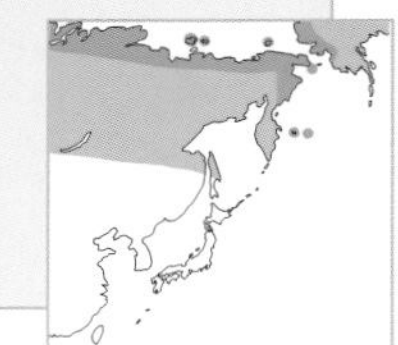

Tengmalm's Owl

Aegolius funereus 24–26cm

Shinya Mori

Adult. January, Hokkaido

Other name Boreal Owl
Description Small, round-headed owl with no ear-tufts. Has a white facial disc with black border, yellow eyes and bill, greyish-brown crown and back with white spots and white underparts with bold brown streaks.
Voice Rapidly repeated and whistled *koo-koo-koo.*
Similar species Little Owl is browner and has an incomplete facial disc.

Range Breeds across northern Eurasia from Europe east through Siberia to the Kamchatka Peninsula, south to north-eastern China, Ussuriland and Sakhalin; also in North America.
Status in Japan Rare resident in coniferous forests on Hokkaido.

Blakiston's Fish Owl

Ketupa blakistoni 63–65cm

Description Very large owl. Greyish-brown above with dark-brown streaks, wings barred buff and brown, yellow eyes and long, broad ear-tufts, buff underparts finely streaked dark-brown.
Voice Male's territorial call is a low, loud far-carrying hooting *buoh-buoh* followed immediately by the female's *woooh*.
Similar species Eurasian Eagle Owl is browner, has black-bordered facial disc with orange eyes, narrower ear-tufts and heavily streaked on the upper-breast.

Kazuyasu Kisaichi

Adult. July, Hokkaido

Range Restricted to far-eastern Asia from Sakhalin, south-eastern Russia south to eastern Hokkaido.
Status in Japan Rare resident in mixed forests along rivers in eastern Hokkaido. Feeds mainly on fish but also takes small rodents in winter.

Kazuyasu Kisaichi

Adult. September, Hokkaido

Manabu Totsuka

Adult. March, Hokkaido

Ural Owl

Strix uralensis 50cm

Tokio Sugiyama

Adult *momiyamae*. April, Aichi

Description Large, round-headed owl with a long barred tail. Has dark brown eyes, yellow bill, mottled dark-brown upperparts and pale underparts with heavy dark streaks. Four races occur in Japan; *japonica* on Hokkaido is pale greyish-white.
Voice Muffled far-carrying hooting; *wo-ho* (silence) *woro-hoho ho voo-hoo*.
Similar species Tawny Owl is much smaller and has a shorter tail.

Akira Yamamoto

Fledgling *momoyimae*. May, Aichi

Range Resident across Eurasia from northern Europe east through Mongolia to far-eastern Russia, north-eastern China and Korea.
Status in Japan Common resident in lowland woodlands from Kyushu northwards. Often found nesting at shrines and temples in foothills where large trees have been preserved.

Mitsuo Imai

Adult *japonica*. January, Hokkaido

Tawny Owl

Strix aluco 38cm

Description Crow-sized, round-headed owl with short tail. Dark-brown eyes, mottled dark-brown back and pale underparts with heavy dark streaks.
Voice Deep, hooting *hoo* (silence) *oo-hoo-hoo*.
Similar species Ural Owl is larger and longer-tailed.

Range Breeds across temperate zone of Eurasia from Europe east to eastern China, Taiwan and Korea. Uncommon resident in Korea and inhabits woodlands and mountain forests.
Status in Japan No records.

Tomi Muukkonen

Adult. February, Finland

Great Grey Owl

Strix nebulosa 64–70cm

Description Unmistakable large owl with heavily-ringed facial discs. Has yellow eyes and bill, back mottled greyish and dark-brown, pale underparts with dark-brown streaks.
Voice Loud and muffled *hoo-hoo-hoo*.

Range Breeds in coniferous forests across northern Eurasia and northern North America. In north-east Asia, breeds from Sakhalin north to Arctic Circle. Sedentary, but irruptions south do occur. Chiefly nocturnal. Hunts over forest clearings and grassy fields.
Status in Japan No records.

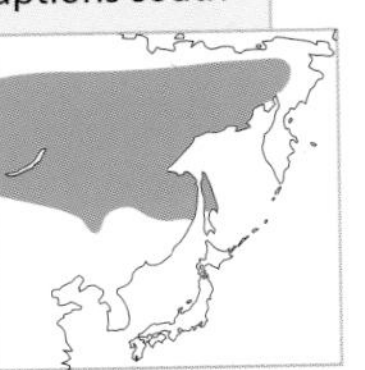

Tomi Muukkonen

Adult. January, Finland.

Eurasian Pygmy Owl

Glaucidium passerinum 16–18cm

Tomi Muukkonen

Adult. December, Lappeenranta, Finland

Description Small, sparrow-sized owl with rounded-head and no ear-tufts. Has yellow eyes and bill, brown upperparts with white spots, white underparts with brown streaks. Often active at dawn and dusk and sits on top of spruce trees, wags tail sideways.
Voice Call is a repeated series of mellow whistles.

Range Breeds in northern Eurasia from central Europe east to Ussuriland, Sakhalin and Heilongjiang province. Mainly sedentary, but nomadic in some areas.
Status in Japan No records.

Little Owl

Athene noctua 23cm

Shin-hwan Kim

Adult. November, Sosan, South Korea

Description Small, round-headed owl with no ear-tufts. Has a dark-brown crown with white streaks, yellow eyes and bill, upper-parts brown with white spots, underparts white with bold brown streaks. More diurnal than other small owls and often seen in open places. Bobs up and down when agitated.
Similar species Tengmalm's Owl is greyer and has black-bordered white facial disc.

Range Widely distributed in temperate Eurasia from Europe east to eastern China and Korea, also in north Africa. In Korea rare resident in the north and a rare winter visitor elsewhere.
Status in Japan No records.

Brown Hawk Owl

Ninox scutulata 29cm

Description Medium-sized hawk-like owl, with long wings and tail. Has a dark-brown head and back, yellow eyes with small white patch over bill and whitish underparts boldly streaked dark-brown. Sexes similar, but male has darker, broader streaks below.

Voice Monotonous and disyllabic *poh-poh, poh-poh*, repeated steadily.

Tokio Sugiyama

Adult ♂ *japonica*. June, Aichi

Range Asiatic. Breeds from Ussuriland south through eastern China and Korea. Northern populations are migratory.

Status in Japan Common summer visitor to wooded areas. Often nests in forested city parks, shrines and temples. Breeds in tree hollows and feeds mainly on large insects. Race *totogo* is resident from the Amami Islands southwards; identical to migratory race *japonica*.

Tadao Shimba

Adult ♀ *japonica*. July, Aichi

Tadao Shimba

Adult ♀ *japonica*. July, Aichi

Eurasian Eagle Owl

Bubo bubo 66cm

Hyuntae Kim

Adult. June, Sosan, South Korea

Description Very large brown owl with orange eyes, dark-bordered brown facial disc and prominent long ear-tufts. Dark-brown upperparts mottled orange-brown; pale brown underparts with bold dark streaks on breast and fine streaks on belly.
Voice Loud, hooting *oo-hoo.*
Similar species Blakiston's Fish Owl has yellow eyes, shorter ear-tufts, greyish-brown back and fine dark streaks on breast.

Range Distributed across temperate Eurasia east to the Okhotsk coast and southern China.
Status in Japan Very rare resident in Hokkaido; vagrant elsewhere. Breeding confirmed for first time in 1994 in northern Hokkaido. Nests in holes on cliffs.

Short-eared Owl

Asio flammeus 37–39cm

Tokio Sugiyama

Adult. January, Aichi

Tokio Sugiyama

Adult. February, Aichi

Description Crow-sized owl with very short ear-tufts and yellow eyes. Has black patches around eyes, mottled brown back and pale-brown underparts with boldly streaked breast and finely streaked belly.
Similar species Long-eared Owl has longer ear-tufts, rufous facial disc and orange-yellow eyes.

Range Distributed across Eurasia, northern North America and southern South America. Mostly migratory in northern hemisphere. Summer visitor to northern part of north-east Asia, winter visitor in south.
Status in Japan Uncommon winter visitor to cultivated fields and open grassy fields. Roosts on the ground in thickets. Partly diurnal.

Long-eared Owl

Asio otus 35–37cm

Description Slender, crow-sized owl with long ear-tufts. Has a dark-bordered rufous facial disc, orange-yellow eyes, mottled greyish-brown upperparts and pale brown underparts with fine dark streaks.
Similar species Short-eared Owl has smaller ear-tufts, yellow eyes and is boldly streaked below on breast.

Range Widely distributed across temperate Eurasia and North America. Summer visitor to northern north-east Asia, winter visitor to Korea.
Status in Japan Uncommon winter visitor to cultivated fields and banks of large rivers from central Honshu southwards. Roosts in small flocks in trees. Breeds locally in northern Honshu and Hokkaido.

Tokio Sugiyama

Adult. December, Aichi

White-throated Needletail

Hirundapus caudacutus 19–21cm WS50–53cm

Description Large dark swift with short blunt tail, pale back, white throat and white undertail-coverts and flanks. Needle-like tail tips are not normally visible in the field.
Similar species Pacific Swift has a white rump, a deeply-forked tail and dark undertail-coverts.

Range Breeds from southeast Siberia east to Sakhalin, south to northern China and Korea; also in the Himalayas. Winters in Australasia.
Status in Japan Uncommon summer visitor to Honshu northwards. Breeds in tree holes in high mountains in Honshu and at lower altitudes in Hokkaido; uncommon elsewhere during autumn migration.

Manabu Totsuka

Adult. June, Gifu.

Manabu Totsuka

Adult. June, Gifu.

Pacific Swift

Apus pacificus 19–20cm WS43–48cm

Norio Yamagata

Adult. July, Shizuoka

Description Large blackish swift with white rump, pale throat and deeply-forked tail. Mottled belly is only visible at close range.
Similar species House Swift is smaller, has shorter wings and a shallow–notched tail.

Tadao Shimba

Adult. July, Shizuoka

Range Breeds from central Siberia east to Kamchatka, south to the Himalayas, southern China and the northern part of south-eastern Asia. Winters in tropical Asia and Australia.
Status in Japan Common summer visitor from Kyushu northwards. Breeds in colonies on cliffs, from rocky shores to high mountains. Common passage migrant in the lowlands.

Tadao Shimba

Adult. July, Shizuoka

House Swift

Apus affinis 13cm WS28cm

Description Small dark swift, with white rump, white throat and shallow–notched tail.
Similar species Pacific Swift is larger, and has longer wings and a deeply forked tail.

Akira Yamamoto

Adult. July, Shizuoka

Range Breeds in south-eastern Asia north to southern China and Taiwan.
Status in Japan Uncommon resident near human habitation from central Honshu southwards. Breeds in colonies and nests on artificial structures such as concrete buildings and bridges; often uses old nests of Asian House Martin. Range has expanded northwards since the 1980s.

Shuichi Tsuno

Adult. April, Wakayama

Shuichi Tsuno

Adult. April, Wakayama

Common Kingfisher

Alcedo atthis 17cm

saw in Karatsu 2008

Akira Yamamoto

Adult ♀ February, Aichi

Description Small kingfisher with a light bluish-green head, greenish-blue back with pale blue centre and blue tail. Also has rusty-red ear-coverts, white neck stripe and rufous underparts, black bill and reddish legs. Sexes similar, but female has reddish base to lower mandible.
Voice Sharp and metallic *tchii*.
Similar species Only small kingfisher in north-eastern Asia.

Tokio Sugiyama

Adult ♀. February, Aichi

Range Breeds across temperate Eurasia from Europe east to the Okhotsk coast and Sakhalin, south to India and south-eastern Asia. Northern populations are migratory and winter south to northern Africa and Indonesia.
Status in Japan Common resident on rivers and ponds throughout Japan. Northern population moves south in winter. Frequently observed on small ponds in city parks and one of the most popular wild birds amongst Japanese people.

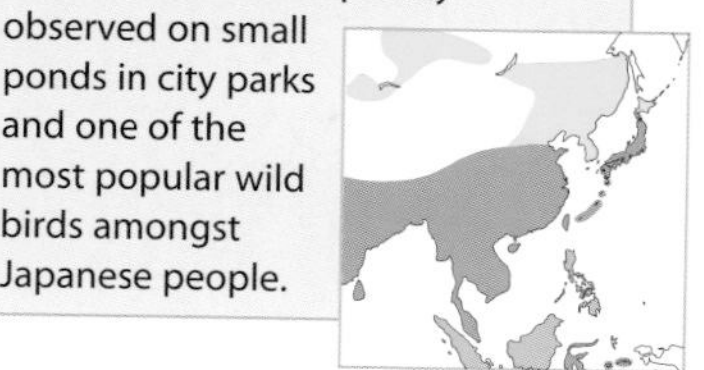

Tadao Shimba

Juvenile. December, Shizuoka

Ruddy Kingfisher

Halcyon coromanda 27cm

Description Large, reddish kingfisher. Has a large red bill, rufous upperparts with purple gloss on wings, yellowish-red underparts and red legs. In flight shows pale bluish rump. Two races in Japan; *bangsi* on Ryukyu Islands has purplish-red back.
Voice Descending clear whistle *pyuu-puy-puy-pu-pu-pu-urrr.*

Kazuyasu Kisaichi

Adult *major*. June, Gifu

Range Breeds from the Himalayas east to south-eastern Asia, north to north-eastern China, Korea and Japan. Northern populations are migratory and winter in tropical Asia.
Status in Japan
Uncommon summer visitor to wooded areas near streams. Fairly common on the Ryukyu Islands.

Manabu Totsuka

Adult *bangsi*. July, Miyako-jima

Akira Yamamoto

Adult *bangsi*. July, Miyako-jima

Black-capped Kingfisher

Halcyon pileata 28cm

Tokio Sugiyama

Adult. May, Tsushima

Description Unmistakable. Medium-sized kingfisher with black cap, white collar and large red bill. Also has a deep shining-blue back and wings with conspicuous large white patch at the base of primaries, white throat and breast and rufous belly.

Norio Yamagata

Adult. May, Tsushima

Range Asiatic. Breeds from India east to eastern China and Korea. Northern populations are migratory and winter south to south-eastern Asia.
Status in Japan Rare passage migrant, mostly observed in spring. Regularly seen on Tsushima and Hegura-jima in May. Coastal areas, cultivated fields and ponds.

Tokio Sugiyama

Adult. May, Tsushima

Crested Kingfisher

Megaceryle lugubris 38cm

Description Large kingfisher with strongly barred black-and-white upperparts, a large shaggy crest, white underparts with a dark breast-band and a yellow-tipped dark bill. Sexes similar, but male has rufous breast-band; female has rufous underwing-coverts.
Voice Loud and sharp repeated note *kak*. Often calls in flight.

Tokio Sugiyama

Adult ♂. February, Aichi

Range Breeds from the Himalayas east to eastern China and Japan.
Status in Japan Uncommon resident from Kyushu to Hokkaido. Inhabits rivers, streams and mountain lakes. Occasionally visits coastal areas in winter. Very shy.

Akira Yamamoto

Adult ♀. February, Aichi

Tokio Sugiyama

Adult ♂. February, Aichi

Grey Nightjar

Caprimulgus indicus 29cm

Akira Yamamoto

Adult ♂. May, Hegura-jima

Description Well-camouflaged nocturnal bird. Has dark greyish-brown upperparts mottled with grey and black spots and streaks. Male has white patch on throat and white spots on four outer primary tips and on outer tail feathers.
Voice Monotonous long and high pitched *kek-kek-kek-kek-kek*; often calls at dusk and dawn.
Similar species The only nightjar in north-east Asia.

Range Breeds from India east to eastern China, Korea and Japan, north to Ussuriland. Northern populations winter in southern Asia.
Status in Japan Uncommon summer visitor to open woodlands from Kyushu northwards. Nests on ground and feeds on insects in flight. Occurs in city parks and gardens on migration. Appears to be declining.

Dollarbird

Eurystomus orientalis 29.5cm

Tadao Shimba

Adult. July, Okayama

Tokio Sugiyama

Adult. June, Gifu

Description Conspicuous large, stocky, bird with dark-green body and broad red bill. Also has a dark purplish-brown head, shiny-blue throat and long, broad wings with distinctive white patches at the base of primaries.
Voice Loud, harsh frog-like croaks *kak*.

Range Widespread in Asia from India to Korea, northern China and the Amur Basin.
Status in Japan Uncommon summer visitor to wooded areas in foothills from Honshu to Kyushu. Declining, possibly due to the loss of suitable nesting holes. A few villages in Okayama have successfully increased breeding population by providing nest boxes.

Eurasian Wryneck

Jynx torquilla 17–18cm

Description Small, long-tailed woodpecker with cryptic plumage. Mottled grey and brown above with pale underparts. Crown and mantle pale grey, dark stripe from eye down neck-side, central dark stripe from crown along back, and has brown bars on throat and belly. Sexes alike. Feeds mainly on ants.
Voice Vocal in breeding season but rather quiet at other times. Loud and harsh *kwee-kwee-kwee*.

Tadao Shimba

Adult. January, Nara

Range Breeds across Eurasia from Europe east to Sakhalin and northern China. Winters mainly in central Africa, India and southern China. Rare winter visitor to Korea.
Status in Japan Uncommon resident. Breeds in forest clearings in northern Japan and winters from Honshu southwards. Occurs along riverbanks and in reed beds in winter; elusive and hard to find.

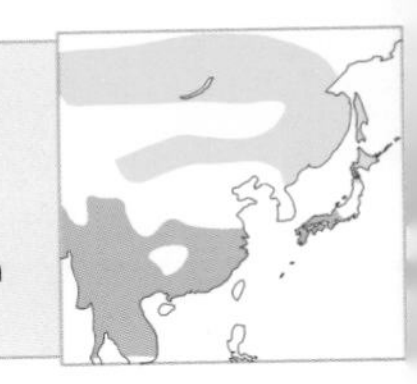

Grey-capped Pygmy Woodpecker

Dendrocopos canicapillus 14–16cm

Description Small woodpecker with grey cap and dark band through ear-coverts down sides of neck. Back and wings black with distinctive white patches on upperwing-coverts, black tail with white bars, pale underparts with dark streaks. Male has red spot on sides of nape, visible at close range.
Voice Call is a series of squeaking notes, *kweek-kweek-kweek,* also soft *chip-chip*.
Similar species Separated from Japanese Pygmy Woodpecker by darker back, white wing patch and white lower back and rump.

Range Resident in Asia from lower Himalayas east to eastern China, Taiwan and south-eastern Asia, north to Ussuriland, north-eastern China and Korea. Woodlands.
Status in Japan No records.

Ran Schols

Adult. May, Beidaihe, China

Japanese Pygmy Woodpecker

Dendrocopos kizuki 15cm

Mitsuo Imai

Adult *ijimae*. December, Hokkaido

Tadao Shimba

Adult *matsudairai*. June, Miyake-jima

Description Small greyish-brown woodpecker with black-and-white barred back; greyish-brown head with white patch above eye, white throat, moustache and sides of neck. Male has small red spot above eye; often difficult to see. Nine races in Japan; northern races are paler and southern races darkest; *ijimae* in Hokkaido is palest and largest.
Voice Call is a distinctive buzzing and grating *tzeeh-tzeeh*, also sharp and rapidly repeated *kik-kik-kik-kik*.
Similar species Lesser Spotted Woodpecker has a more pied appearance and a different call. See also Grey-capped Pygmy Woodpecker.

Range Sakhalin, Ussuriland south to northeast China, Korea and Japan.
Status in Japan Common resident in lowland forests and often common in city parks.

Tadao Shimba

Adult *seebohmi*. February, Aichi

Great Spotted Woodpecker

Dendrocopos major 23cm

Description Medium-sized, black-and-white woodpecker with conspicuous white shoulder patches and pale underparts with red vent and belly. Male has black crown with red nape, while female has entirely black crown. Three races recorded in Japan; *japonicus* in Hokkaido has larger white patches on back than *hondoensis* in Honshu. Vagrant *brevirostris* has thicker bill and larger white patches on back than other two races.
Voice Sharp and loud *kik*.
Similar species White-backed Woodpecker is larger, has distinctive black streaks on underparts.

Tadao Shimba

Adult ♀ *japonicus*. January, Hokkaido

Yoshihito Goto

Adult ♀ *brevirostris*. October, Hegura-jima

Range Widely distributed across Eurasia from Europe east to the Okhotsk coast, Kamchatka and Sakhalin, south to north-eastern China and Korea.
Status in Japan Common resident in deciduous forests in Honshu and Hokkaido.

Akira Yamamoto

Adult ♂ *hondoensis*. May, Aichi

Akira Yamamoto

Adult ♀ *hondoensis*. May, Aichi

Rufous-bellied Woodpecker

Dendrocopos hyperythrus 24cm

Ran Schols

Adult ♂. May, Beidaihe, China

Description Medium-sized woodpecker with a distinctive rufous neck and underparts, white face, white bars on dark upperparts and red undertail-coverts. Male has red crown, female has black crown with white streaks.

Range Breeds in southern Asia, northern south-east Asia, southern Ussuriland, north-eastern China and northern Korea. Northern populations are migratory and winter south to eastern China.
Status in Japan Vagrant. One record in May 2005 on Hegura-jima.

Lesser Spotted Woodpecker

Dendrocopos minor 16cm

Kazuyasu Kisaichi

Adult ♀. June, Hokkaido

Description Small and compact, essentially black-and-white woodpecker with barred back and whitish underparts. Male has red crown, white on female.
Voice Call is a sharp *kik*.
Similar species Japanese Pygmy Woodpecker is browner, has extensive brown on ear-coverts and a greyish-brown crown.

Kazuyasu Kisaichi

Adult ♂. June, Hokkaido

Range Widely distributed across Eurasia from Europe east to the Kamchatka peninsula, Sakhalin and northern China.
Status in Japan Uncommon resident in eastern Hokkaido. Only a few records from elsewhere. Prefers open lowland forests. Sometimes observed in parks, river-banks and coastal areas.

White-backed Woodpecker

Dendrocopos leucotos 28cm

Description Medium-large, black-and-white woodpecker. Black wings with white barring and no distinctive white patches on back. Underparts white with black streaks, belly pinkish. Rump and back white. Male has entirely red crown, female has black crown. Four races in Japan; southern races are darker, with *owstoni* on Amami Islands the darkest.
Voice Loud and sharp *kik*.
Similar species Great Spotted Woodpecker is smaller, has extensive white shoulder patches and lacks streaks on underparts.

Range Widely distributed across Eurasia from western Norway and the Balkans east to the Sea of Okhotsk, Kamchatka Peninsula and Sakhalin, south to north-eastern China and Korea.
Status in Japan
Uncommon resident in forests from Kyushu northward and on the Amami Islands.

Tadao Shimba

Adult ♂ *subcirris*. January, Hokkaido

Tadao Shimba

Adult ♀ *subcirris*. January, Hokkaido

Kazuyasu Kisaichi

Adult ♂ *owstoni*. May, Amami-Oshima

Black Woodpecker

Dryocopus martius 46cm

Tadao Shimba

Adult ♂. January, Hokkaido

Description Large, entirely black woodpecker with pale yellow bill and eye. Male has an extensive red crown and female has a red patch on rear of crown.
Voice Loud, plaintive whistle *pyuuuuh,* also rolling *krri-krri-krri-krri.*

Range Widely distributed across Eurasia from Europe east to the Sea of Okhotsk, Kamchatka Peninsula and Sakhalin, south to northern China and Korea.
Status in Japan Uncommon resident in deciduous and mixed forests in Hokkaido. Mostly feeds on lower tree trunks and on the ground. Relict population in northern Honshu is critically endangered.

Yoshihito Goto

Adult ♂. May, Hokkaido

Kazuyasu Kisaichi

Adult ♀. May, Hokkaido

Japanese Woodpecker

Picus awokera 29cm

Description Medium-sized woodpecker with green back and yellow rump. Greyish face with black lores and red moustache, neck and breast grey, heavily barred pale belly. Male has red crown and female has grey crown with red nape patch.
Voice Quite vocal in breeding season. Loud, sharp whistle *pee-yoh,* often repeated, and also sharp *ket ket*.
Similar species Grey-headed Woodpecker lacks heavy barring on underparts and its range does not overlap.

Range Endemic to Japan.
Status in Japan Common resident from Kyushu north to Honshu. Inhabits evergreen and deciduous forests. Often visits city parks and wooded urban areas in winter.

Tadao Shimba

Adult ♂. February, Nagano

Akira Yamamoto

Adult ♂. June, Aichi

Akira Yamamoto

Adult ♀. June, Aichi

Grey-headed Woodpecker

Picus canus 29–30cm

Kazuyasu Kisaichi

Adult ♂. December, Hokkaido

Description Medium-sized woodpecker with green back, pale greenish-grey underparts and yellow rump; greyish head and face with dark lores and dark moustache. Male has bright red forehead and female has plain grey crown.
Voice Descending *pew-pew-pew,* and also sharp *kik-kik.*
Similar species Japanese Woodpecker has heavily barred belly and range does not overlap.

Mitsuo Imai

Adult ♀. December, Hokkaido

Range Widely distributed across Eurasia from Europe east to south-eastern Russia and Sakhalin, south to Korea, eastern China and northern part of south-eastern Asia.
Status in Japan Uncommon resident in mixed and deciduous forests on Hokkaido. A few records from Honshu.

Three-toed Woodpecker

Picoides tridactylus 20–24cm

Tomi Muukkonen

Adult ♂. May, Lappeenranta, Finland

Description Medium-sized woodpecker with black and white striped head, and white underparts with dark streaked flanks. Black wings with white bars on flight feathers, white back, black tail. Male has yellow crown, female has dark crown with white streaks.

Range Breeds in coniferous forests across northern Eurasia east to Kamchatka, south to Ussuriland, Sakhalin, northeast China and northern Korea; also in western China.
Status in Japan Very rare resident in coniferous forests in Hokkaido.

Okinawa Woodpecker

Sapheopipo noguchii 31cm

Description Medium-sized, dark-brown woodpecker with a pale yellow bill, pale-brown face and throat, deep red-brown back and belly and small white spots on primaries. Male has entire crown red with dark streaks, while female has a dark crown.
Voice Call is *kuk or kyak,* softer than that of Great Spotted Woodpecker.

Kazuyasu Kisaichi

Adult ♀. May, Okinawa-Hontou

Range Endemic to Okinawa-Hontou.
Status in Japan
Uncommon resident in northern part of the island. Inhabits undisturbed evergreen subtropical forests. Threatened.

Kazuyasu Kisaichi

Adult ♂. May, Okinawa-Hontou

Eurasian Hoopoe

Upupa epops 26cm

Tokio Sugiyama

Adult. May, Hegura-jima

Description Unmistakable. Orange-brown body with black and white wings and tail. Distinctive black-tipped crest and long, downcurved bill. Flight buoyant; spreads fan-shaped crest when alert.
Voice Far-carrying hollow *poo-poo-poo*.

Range Widely distributed in Eurasia and Africa. In north-eastern Asia, common summer visitor from Amur Basin and Ussuriland south to Korea.
Status in Japan Uncommon visitor to grassy and cultivated fields. Most often seen in western Japan; mainly in March and April. A few breeding records from Honshu.

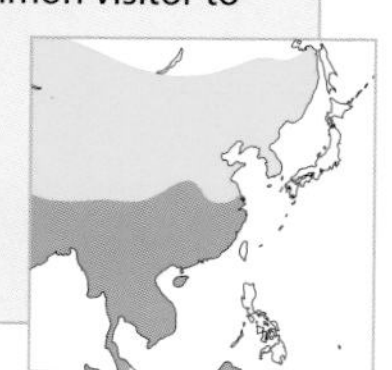

Fairy Pitta

Pitta nympha 18cm

Mitsuo Imai

Adult. May, Hegura-jima

Description Colourful plump bird with a very short tail, reddish-brown crown with black centre, yellowish supercilium, conspicuous broad black eye-stripe, emerald-green back with bright-blue lesser–coverts, and bright-blue rump.
Voice Call is a loud whistle *churr-ru churr-re.*

Range Breeds locally in South Korea, south-east China, Taiwan and southern Japan. Migratory, wintering mainly in Borneo.
Status in Japan Uncommon summer visitor to dense evergreen forests in southern Japan. Skulks secretively on the forest floor and is normally difficult to observe.

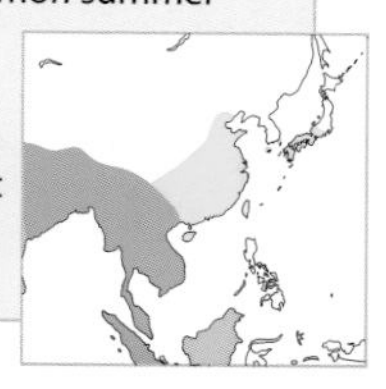

Bimaculated Lark

Melanocorypha bimaculata 17cm

Description Large, sandy-brown lark with a heavy bill. Has fine streaks on head and back, pale eye-ring and supercilium, whitish underparts with conspicuous black patches on sides of neck and white tips to tail feathers.
Similar species Eurasian Skylark has short crest, longer tail and thinner bill and lacks black neck-patch. Both Greater Short-toed Lark and Asian Short-toed Lark are smaller.

Range Breeds from Turkey east to central Asia. Winters mainly from north-eastern Africa to India.
Status in Japan Vagrant with several wintering records. Occurs in cultivated and grassy fields.

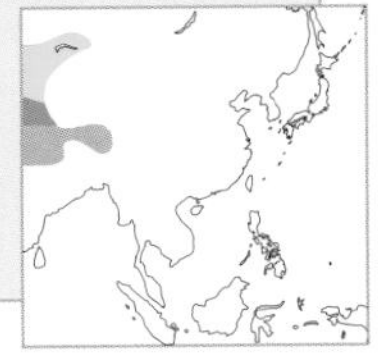

Akira Yamamoto

Adult. February, Aichi

Greater Short-toed Lark

Calandrella brachydactyla 14cm

Description Small pale-brown lark with short, pointed bill. Has fine dark streaks on crown and back, buff-coloured eye-ring and supercilium and dark eye-stripe behind eye. Whitish underparts with buff-coloured wash on breast and flanks.
Voice Call is a short *chip-chip* or *junh-junh*.
Similar species Asian Short-toed Lark has sandy-brown back with fine dark streaks on breast and shorter tertials (primary projection beyond tertials).

Range Breeds from Europe and northern Africa east through central Asia to Mongolia and northern China. Winters in Africa and in western and southern Asia.
Status in Japan Rare passage migrant, mainly on offshore islands in Sea of Japan. Seen on cultivated and grassy fields.

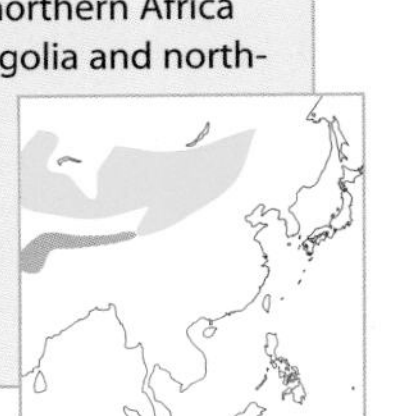

Yoshio Osawa

Adult. April, Tobishima

Asian Short-toed Lark

Calandrella cheleensis 14cm

Akira Yamamoto

Adult. December, Aichi

Description Small, sandy-brown lark with small, stout, pale-yellowish bill. Has pale underparts with fine dark streaks on breast. Primaries project beyond tertials.
Voice Soft *purrh-purrh.*
Similar species Greater Short-toed Lark has browner back, longer tertials and longer bill.

Range Breeds in Transbaikalia, north-eastern Mongolia and northern China. Winters south to southern Mongolia and north-eastern China.
Status in Japan Rare passage and winter visitor seen on cultivated and grassy fields.

Crested Lark

Galerida cristata 17cm

Hyun-tae Kim

Adult. April, Sosan, South Korea

Description Large lark with distinctive pointed crest on crown. Sandy-brown above, streaked darker. Has a long bill, whitish throat and underparts with streaked breast and a short tail with rusty-coloured sides.
Voice A melodious *dooee,* usually calls on taking off.
Similar species Separated from Eurasian Skylark by pointed crest and flight pattern.

Range Breeds across Eurasia from Europe east through central Asia to central China and Korea; also in northern Africa.
Status in Japan No records.

Eurasian Skylark

Alauda arvensis 17cm

Description One of the commonest birds in Japan. Has brown upperparts with finely streaked head and nape and heavy dark streaks on back; short crest, white supercilium, yellowish-brown bill, whitish underparts with fine streaks on breast and pinkish legs. In flight shows white trailing edges to secondaries and white outer-tail feathers.
Voice Song is a continuous stream of musical trills given mostly high up in the air. Call is loud *byurrh*.
Similar species Oriental Skylark *A. gulgula* is smaller, with longer tertials, shorter tail and lacks white trailing edge to secondaries.

Akira Yamamoto

Adult. June, Aichi

Range Widespread in temperate Eurasia from Europe east through Siberia to Kamchatka and Sakhalin, south to northern China and Korea. Northern populations are migratory.
Status in Japan Common resident in cultivated fields, river-banks and grassy fields from Kyushu northward. Summer visitor to Hokkaido and winter visitor to Ryukyu Islands.

Akira Yamamoto

Adult. June, Aichi

Tadao Shimba

Adult. March, Aichi

Horned Lark

Eremophila alpestris 16–17cm

Kazuyasu Kisaichi

Adult non-breeding. February, Saitama

Description Pale brownish lark with characteristic facial pattern. Yellow forehead, supercilium and throat contrast with black stripe on side of head and cheeks, broad black breast-band, whitish underparts and black legs. In breeding plumage shows small black 'horns' on head.

Range Breeds in the Arctic and in dry mountains and steppe of Eurasia, also in North America. Northern populations are migratory.

Status in Japan Rare winter visitor to fields and coastal areas.

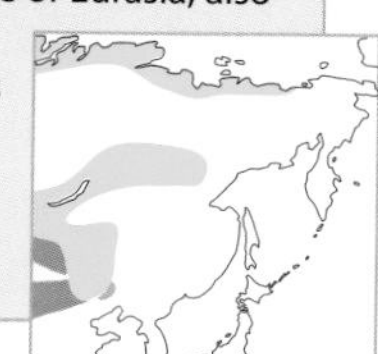

Sand Martin

Riparia riparia 12–13cm

Tadao Shimba

Adult. August, Ibaragi

Akira Yamamoto

Juvenile. September, Aichi

Description Small brown swallow with brown breast-band, white throat and underparts.

Similar species Asian House Martin has purplish-black back and conspicuous white rump and uppertail-coverts.

Range Breeds across Eurasia from Europe east to Kamchatka, south to northern Africa, Asia Minor, northern India and eastern China, also in North America. Winters south to tropical regions. Uncommon passage migrant in Korea.

Status in Japan Locally common summer visitor to Hokkaido. Breeds in colonies in earth banks. Common passage migrant elsewhere. Common in rice fields in summer and early autumn.

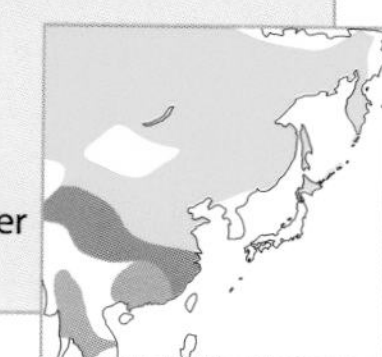

Asian House Martin

Delichon dasypus 13cm

Description Small swallow, glossy purplish-black above, whitish below with pale buff-coloured wash on breast and flanks. White rump and short, shallow-forked tail.
Similar species House Swift is larger, and has longer wings and dark underparts. Northern House Martin *D. urbicum,* a rare visitor to Japan, has whiter belly, larger white patch on uppertail-coverts and slightly deeper–forked tail.

Range Southern and eastern Asia. Breeds from Indian subcontinent east to south-east Asia, north to Ussuriland and Sakhalin. Winters south to southern Asia.
Status in Japan Common summer visitor from Kyushu northwards. Breeds in colonies on cliffs; also on artificial structures such as buildings and bridges. Many winter in southern Japan.

Tadao Shimba

Adult. May, Gifu

Tokio Sugiyama

Adult. June, Aichi

Manabu Totsuka

Adults. July, Hokkaido

Barn Swallow

Hirundo rustica 17–18cm

Tadao Shimba

Adult. September, Aichi

Shuichi Tsuno

Adult. July, Wakayama

Description Glossy blue-black above, white underparts with black breast-band and reddish-brown forehead and throat. Deeply-forked tail with white spots. Juvenile is browner and lacks long outer-tail streamers.
Similar species See Pacific Swallow.

Range Widespread throughout Eurasia, also in northern Africa and North America. Winters mainly in tropical regions. In north-east Asia a common summer visitor except in northern regions.
Status in Japan Most common swallow and well adapted to human habitation. Frequently nests under the eaves of houses and normally welcomed by local people. Abundant summer visitor throughout and a few winter in southern Japan. Forms large flocks and roosts in reed beds in late summer before south-bound migration.

Akira Yamamoto

Juveniles. July, Aichi

Pacific Swallow

Hirundo tahitica 13cm

Description Has a glossy blue-black back, greyish-white underparts with dusky, scaly marks on undertail-coverts. Reddish-brown forehead and throat. Juvenile is duller.
Similar species Barn Swallow is larger, has dark breast-band and white belly and undertail-coverts.

Tadao Shimba

Adult. September, Okinawa-Hontou

Range Breeds in southern Asia from southern India east to South-east Asia, north to Taiwan and the Ryukyu Islands.
Status in Japan Uncommon resident on Ryukyu Islands. Frequently seen perched on telephone wires around coastal villages. Nests in colonies under the eaves of houses or on bridges.

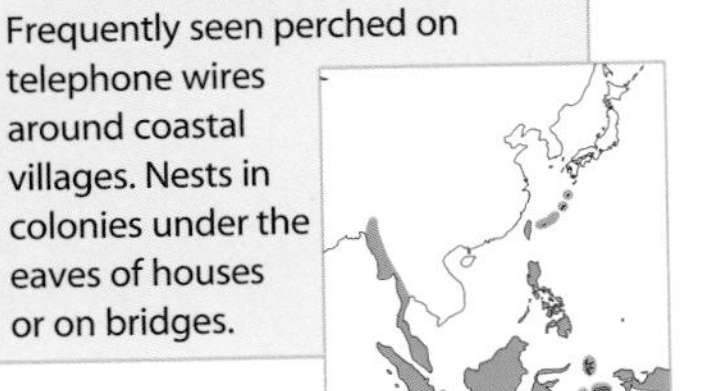

Kiyoto Ogata

Adult. May, Okinawa-Hontou

Tadao Shimba

Juvenile. July, Miyako-jima

Red-rumped Swallow

Hirundo daurica 18–19cm

Tokio Sugiyama

Adult. July, Aichi

Description Medium-sized swallow with reddish-brown rump and deeply-forked tail, underparts buff-coloured with distinctive black streaks.

Similar species Barn Swallow has white underparts and dark rump.

Range Breeds in Eurasia from southern Europe east through Asia Minor to eastern China, north to Ussuriland and lower Amur Basin; also in Africa. Northern populations are migratory and winter south to Africa and southern Asia.

Status in Japan Uncommon summer visitor to southern Japan, rare in the north. Breeds in colonies often near human habitations and nests on artificial structures such as buildings and bridges.

Shuichi Tsuno

Adult. July, Wakayama

Shuichi Tsuno

Adult. July, Wakayama

Forest Wagtail

Dendronanthus indicus 15cm

Description Olive-brown head with white supercilium, olive-brown back and whitish underparts with distinctive black double breast-bands. Dark wings with two white wing-bars; legs are pinkish. Wags tail and body sideways in a unique fashion.
Voice Song is a loud disyllabic note, similar to that of Great Tit. Call is a thin, high-pitched *tzeeeh*.

Range Breeds in eastern Asia from southern Ussuriland and north-eastern China, south to eastern and central China. Winters south to south-eastern Asia.
Status in Japan Rare passage migrant, mainly to western Japan, most often seen on offshore islands in Sea of Japan. Breeding records from Fukuoka and Shimane. Usually seen in wooded areas.

Tadao Shimba

January. Nara

Teruaki Ishi

January. Nara

Akira Yamamoto

May. Tsushima

White Wagtail

Motacilla alba 21cm

Description Most common black and white wagtail. Six races recorded in Japan. Most common race *lugens;* male has black crown, neck and back, white face with black eye-stripe and black breast. Female and non-breeding plumage greyer. In flight shows largely white wings. Male *leucopsis* is similar to *lugens,* but lacks black eye-stripe. Male *ocularis* has black eye-stripe, grey back and dark flight feathers. Male *baicalensis* similar to *ocularis,* but lacks black eye-stripe. Males of *alboides* and *personata* have black heads with white foreheads and ear-coverts.
Voice Song is a twittering with call-like notes, call is a metallic disyllabic *tsik-tsik.*
Similar species Japanese Wagtail has black face and different flight call.

Range Widespread throughout Eurasia and winters south to northern Africa, Indian subcontinent and south-eastern Asia. In north-eastern Asia, *ocularis* breeds in north-eastern Siberia, south to 60°N; *leucopsis* breeds from Ussuriland south to Korea; *lugens* mainly in Ussuriland, Sakhalin and Japan.
Status in Japan Race *lugens* is an abundant resident from Honshu northwards. Moves to warmer areas in winter and occurs in a wide variety of open habitats, even in large cities. Race *leucopsis* is an uncommon passage and winter visitor to southern Japan; *ocularis* is a rare passage migrant mainly to offshore islands around Sea of Japan; other races are vagrants.

Tadao Shimba

Adult ♂ breeding *lugens*. May, Hegura-jima

Shuichi Tsuno

Adult ♂ non-breeding *lugens*. Oct., Wakayama

Akira Yamamoto

Adult ♂ breeding *leucopsis*. May, Miyako-jima

Tadao Shimba

Adult ♂ breeding *ocularis*. May, Hegura-jima

Himaru Iozawa

Adult ♂ breeding *personata*. March, Yonaguni-jima

Tadao Shimba

First-winter *lugens*. October, Aichi

Japanese Wagtail

seen often near our quarters

Motacilla grandis 18-21cm

Description Black upperparts and breast contrast with white underparts. Has prominent white supercilium and white chin, white wings with black primaries and black leading edges to wing coverts. Sexes similar, but female normally has sooty-black back.
Voice Song is a variety of melodious twittering mixed with call-note, delivered from a high exposed perch. Flight call is a short and harsh disyllabic *tzit-tzit*.
Similar species White Wagtail has white face and different flight call.

Mitsuo Imai

Adult ♂. February, Ibaragi

Range Endemic to Japan and also occurs in southern Korea, particularly in winter.
Status in Japan Common resident from Kyushu northwards. Breeds near running water and inhabits relatively lower altitudes than Grey Wagtail in summer. In winter, occurs around rivers and lakeshores. Vagrant to the Ryukyu Islands.

Akira Yamamoto

Adult ♂. November, Aichi

Tadao Shimba

Adult ♂. November, Aichi

Yellow Wagtail

Motacilla flava 16.5cm

Tadao Shimba

Adult breeding *taivana*. July, Hokkaido

Mitsuo Imai

Adult breeding *taivana*. May, Yonaguni-jima

Tadao Shimba

Adult ♂ breeding *simillima*. May, Tsushima

Description Four races recorded in Japan; *taivana* and *simillima* most often observed. Race *taivana* has greenish-olive back and dark wings with buff-coloured edges to wing-coverts and tertials, yellow supercilium, greenish-olive eye-stripe and yellow underparts; *simillima* has bluish-grey head with white supercilium and white throat; *macronyx* has dark-grey head and lacks supercilium; *plexa* has dark-grey head with supercilium from rear of eye. Females generally duller than males. Juveniles have whitish underparts, prominent white supercilium and greyish to olive-brown back.

Voice Song is a series of variable short chirps, call is a sharp and explosive *tzreep*.

Similar species Grey Wagtail has grey back, longer tail, pinkish legs and a more metallic flight call. Citrine Wagtail has a different facial pattern.

Range Widespread in the Palearctic region. Breeds in northern and temperate regions of Eurasia, also in western Alaska. Winters in Africa and southern Asia. In North-east Asia, breeds from north-eastern China northward. Common passage migrant to Korea.

Status in Japan Rare passage migrant mainly to offshore islands in Sea of Japan. Race *taivana* breeds in freshwater swamps in northern Hokkaido and is a common winter visitor to Ryukyu Islands. Occurs on cultivated and grassy fields during migration.

Kazuyasu Kisaichi

Adult ♂ breeding *macronyx*. May, Tsushima

Tadao Shimba

Juvenile. September, Okinawa-Honto

Grey Wagtail

Motacilla cinerea 20cm

Description Long-tailed wagtail with grey upperparts and yellow rump, dark wings with white fringes to tertials, white supercilium, yellow underparts with white flanks and pinkish legs.
Voice Song is a high-pitched and rapidly repeated, *tse-tse-tse-tse,* given from exposed perch. Flight call is a thin, metallic disyllabic *tsit-tsit,* similar to that of White Wagtail but sharper.
Similar species Yellow Wagtail has olive-green back and black legs, shorter tail and a different call.

Range Breeds in Siberia from the Urals east to Kamchatka, south to central and eastern China; also in western and central Europe, northern Africa and the Himalayas. Northern populations are migratory and winter in the Indian subcontinent and south-east Asia.
Status in Japan Common resident; breeds near running water. Moves to warmer areas in winter and a winter visitor to Ryukyu Islands. Inhabits rivers and ponds in winter.

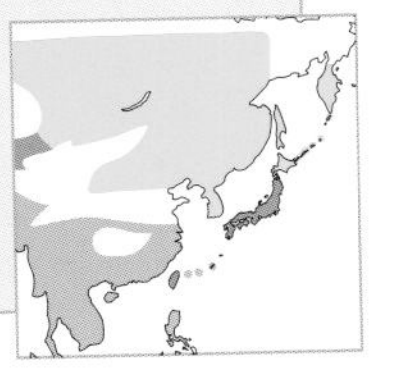

Tadao Shimba

Adult ♂ breeding. May, Nagano

Tadao Shimba

Adult ♂ breeding. May, Nagano

Manabu Totsuka

Adult non-breeding. February, Aichi

Citrine Wagtail

Motacilla citreola 16.5cm

Tokio Sugiyama

Adult ♂ breeding. May, Tsushima

Description Breeding male has distinctive yellow head, breast and belly; dark-grey back and black wings with broad white fringes to wing–coverts and tertials. Female has greyish crown and ear-coverts. Non-breeding male resembles female. Juvenile lacks yellow; grey ear-coverts are white-framed.
Voice Song similar to that of Yellow Wagtail; call is a sharp, explosive *tzreep.*
Similar species Non-breeding adults and juveniles separated from juvenile Yellow Wagtail by grey upperparts and white (not yellow) undertail-coverts.

Yoshio Goto

Adult ♀ breeding. May, Hegura-jima

Range Breeds in northern Eurasia from north-western Siberia east through central Siberia to Transbaikalia and north-western China, south through central Asia to the Himalayas. Winters mainly in southern Asia from Pakistan east to southern China.
Status in Japan Very rare passage migrant; most records from islands in Sea of Japan and the Ryukyu Islands. Found on cultivated and grassy fields.

Norio Yamagata

Adult ♂ breeding. May, Tsushima

Meadow Pipit

Anthus pratensis 14.5cm

Description Olive-brown above with dark streaks, indistinct supercilium, white eye-ring and dark bill with orange lower mandible. Whitish below with bold dark streaks on breast and flanks. Legs pinkish.
Voice Flight call is a loud and shrill *pseet* similar to Buff-bellied Pipit.
Similar species Buff-bellied Pipit has more prominent pale supercilium, and a greyish-brown back with indistinct dark streaks.

Kazuyasu Kisaichi

Adult non-breeding. February, Fukuoka

Range Breeds in northern Europe and north-western Asia east to the Ob River. Winters south to the Mediterranean and south-western Asia.
Status in Japan First recorded in 1997 in Fukuoka. Several subsequent reports from Ryukyu Islands and Hegura-jima. Occurs in cultivated fields and grassy meadows.

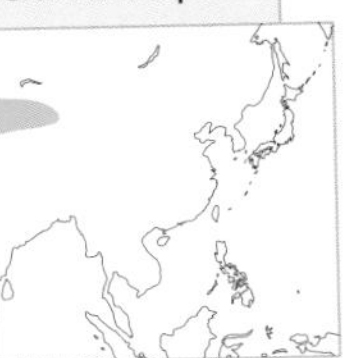

Yoshihito Goto

Adult non-breeding. December, Amami-Oshima

Yoshihito Goto

First-winter. December, Amami-Oshima

Richard's Pipit

Anthus richardi 18cm

Tokio Sugiyama

Adult breeding. May, Tsushima

Description Large, buff-coloured pipit with pale supercilium, large bill and buff-coloured breast with dark streaks. White underparts and long legs with long hind claws. At close range shows triangular-shaped dark centres to median-coverts in fresh adult plumage. Often takes upright posture when alert.

Voice Song is a series of buzzing notes and flight call is a loud and harsh *shreep.*

Similar species Blyth's Pipit is slightly smaller, has a smaller bill, shorter tail and legs, shorter hind claws, narrower breast streaks, an indistinct supercilium and squarer dark centres to median-coverts.

Akira Yamamoto

Adult non-breeding. November, Aichi

Range Breeds in Siberia east to the Okhotsk coast, south to Mongolia and eastern China. Winters south to the Indian subcontinent and south-eastern Asia. Uncommon passage migrant to Korea.

Status in Japan Rare passage migrant mainly in western Japan. Some winter in southern Kyushu and on Ryukyu Islands. Seen on cultivated fields and grassy meadows.

Tadao Shimba

First-winter. January, Aichi

Blyth's Pipit

Anthus godlewskii 17cm

Description Similar to Richard's Pipit, but slightly smaller with shorter tail, legs and hind claws; a relatively small bill and narrow dark streaks on breast. At close range shows square centres to median-coverts in fresh adult plumage.
Voice Flight call is a short and metallic *chip*.

Range Breeds in southern Siberia from the Altai east to Transbaikalia, south to Mongolia and north-western China. Winters mostly in the Indian subcontinent.
Status in Japan Rare passage migrant on cultivated and grassy fields, mainly in western Japan; some winter records from the Ryukyu Islands.

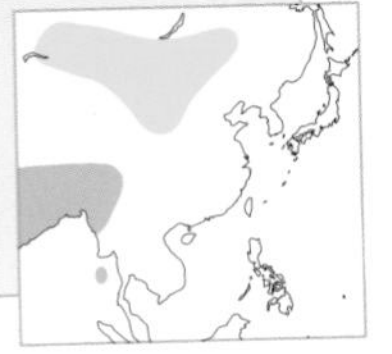

Teruaki Ishii

Adult breeding. May, Hegura-jima

Teruaki Ishii

Adult breeding. May, Hegura-jima

Shigeo Ozawa

Adult breeding. May, Tsushima

Pechora Pipit

Anthus gustavi 14cm

Hyun-tae Kim

Adult breeding. May, Sosan, South Korea

Description Upperparts greyish-brown with heavy black streaks on back and one or two white stripes on the mantle sides. Buffish below with heavily streaked breast and flanks. Secretive and often remains in cover; hard to see.
Voice Often silent when flushed. Call is a short and squeaky bunting-like *tsip*. Song is a series of rapidly repeated short notes.
Similar species Red-throated Pipit has a larger bill, rufous or orange-brown throat in adults and longer tertials (no primary projection); also has different call.

Hyun-tae Kim

Adult breeding. May, Sosan, South Korea

Range Breeds across northern Siberia east to the Bering Sea and Kamchatka, also in north-eastern China (Heilongjiang province) and south-eastern Russia. Winters mainly in the Philippines and Indonesia.
Status in Japan
Rare passage migrant, mainly in western Japan, with small numbers seen annually on the Ryukyu Islands.

Mitsuo Imai

First-winter. October, Hegura-jima

Olive-backed Pipit

Anthus hodgsoni 15cm

Description Upperparts greenish-brown with dark streaks. Has a clear white supercilium and a distinctive white spot on the ear-coverts. Buffish-white underparts with heavy streaks on breast and flanks; pinkish legs with relatively short hind claws. In non-breeding plumage has yellowish edges to median-coverts.
Voice Song contains a variety of melodious trills followed by drawn-out soft whistled *seeh, seeh, seeh*. Often delivered from the top of coniferous trees or in short song flights. Call is a sharp, harsh *bizzt*.
Similar species See Tree Pipit.

Tadao Shimba

Adult breeding. June, Nagano

Range Breeds in the taiga belt of Siberia from the Urals east to the Okhotsk coast and Kamchatka, south to central China and northern Korea. Winters south to the Indian subcontinent and south-eastern Asia.
Status in Japan Common resident; breeds in subalpine coniferous forests and highland meadows from Shikoku northwards. Moves to warmer areas in winter and inhabits open lowland forests and forested city parks.

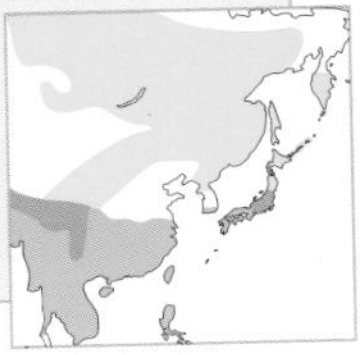

Tadao Shimba

Adult non-breeding. February, Aichi

Tadao Shimba

Adult non-breeding. February, Aichi

Red-throated Pipit

Anthus cervinus 15cm

Tadao Shimba

Adult breeding. April, Ishigaki-jima

Description Upperparts greyish-brown with heavy black streaks on back, rump and uppertail-coverts. In breeding plumage has rusty-red supercilium, throat and breast and buff-coloured underparts with black streaks on breast and flanks. Rusty-red on throat and breast is orange-brown in winter. Juveniles have a pale supercilium and uniform buff-coloured underparts, lacking rusty-red on breast.
Voice Common flight call is a thin, high and metallic *tseep*.
Similar species See Pechora Pipit.

Manabu Totsuka

Adult non-breeding. January, Okinawa-Hontou

Range Breeds on tundra in northern Eurasia from Scandinavia east to the Bering Sea and Kamchatka; also in western Alaska. Winters mainly in Africa, south-east Asia, southern China and Taiwan.
Status in Japan Uncommon passage migrant and winter visitor mainly to western Japan. Occurs in rice fields and grassy fields.

Kenji Takehara

Adult breeding. April, Okinawa-Hontou

Buff-bellied Pipit

Anthus rubescens 16cm

Description Dark greyish-brown above with indistinct streaks, two white wing-bars, pale supercilium and eye-ring; dark bill with yellowish base to lower mandible, pinkish legs. Whitish below with bold, dark streaks on breast and flanks. In breeding plumage has a greyish back and pale reddish-brown underparts.
Voice Call is a loud and shrill *pseet*.
Similar species Meadow Pipit has an olive-brown back with dark streaks, also lacks a distinct pale supercilum.

Tokio Sugiyama

Adult breeding. April, Aichi

Range Breeds in north-eastern Russia south to Lake Baikal and Sakhalin; also in North America. Winters south to south-eastern China, Korea and Japan. Common winter visitor to Korea.
Status in Japan Common winter visitor to rice fields, riverbanks and lake shores. Arrives in September and departs in the first half of April.

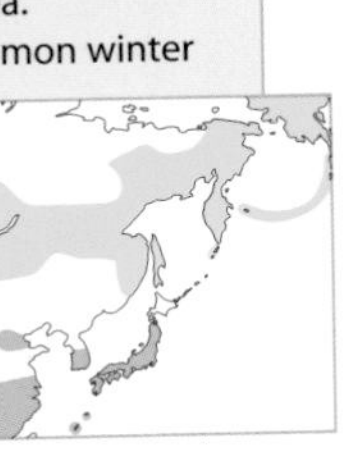

Tadao Shimba

Adult non-breeding. January, Aichi

Tadao Shimba

Adult non-breeding. January, Aichi

Tree Pipit

Anthus trivialis 15cm

Tokio Sugiyama

Adult. May, Hegura-jima

Description Pale olive-brown head and back with dark streaks, buff coloured supercilium, dark bill with pale base to lower mandible. Breast and flanks buff, streaked brown. Legs pinkish, with relatively short hind claws.
Voice Call is a sharp, harsh *bizzt,* similar to Olive-backed Pipit.
Similar species Olive-backed Pipit has greenish-brown back, prominent white supercilium and white and dark spots on ear-coverts.

Range Breeds across Eurasia from Europe east to the Kolyma River in the north and Lake Baikal in the south. Winters mainly in tropical Africa and the Indian subcontinent.
Status in Japan Rare passage migrant seen on cultivated fields and open woodlands. Mainly observed on offshore islands in Sea of Japan in spring.

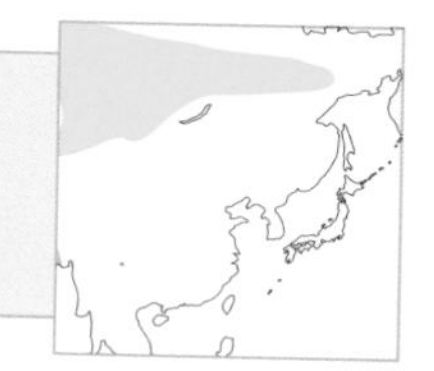

Light-vented Bulbul

Pycnonotus sinensis 18–19cm

Kazuyasu Kisaichi

Adult. May, Okinawa-Hontou

Description Black head with a distinctive large white patch on hind-crown and a small white patch on the ear-coverts. Olive-brown above with bright olive fringes to wings and tail, white throat and pale greyish-brown underparts.
Voice Loud, metallic and variable trills.

Range Resident in the northern part of south-eastern Asia, southern China, Taiwan and the Ryukyu Islands.
Status in Japan Common resident on Okinawa-Hontou and Yaeyama Islands. Wooded urban areas and cultivated fields.

Brown-eared Bulbul

Ixos amaurotis 27–28cm

Description Silvery-grey head and throat with chestnut ear-coverts, greyish-brown back and streaked greyish underparts. Long, dark bill and dark-red legs. Several races in Japan, with the southern races darker.
Voice Loud, sharp and metallic *peet-peet*, also loud screaming *pee-yoh*.

Tadao Shimba

Adult *squameiceps*. August, Chichi-jima

Range Resident in Japan and Korea. The northern population is migratory.
Status in Japan Abundant resident in wooded urban areas and deciduous forests. The northern population moves south in winter. Common in city parks and gardens and often dominates at bird feeders. Flocks of thousands can be seen in autumn.

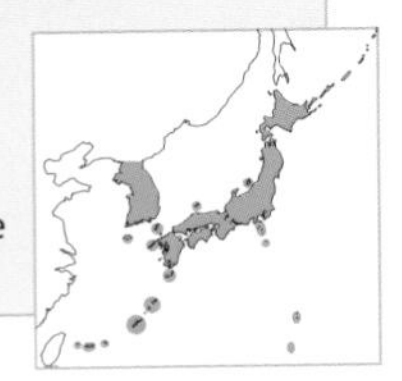

Manabu Totsuka

Flock of *amaurotis*. October, Aichi

Tokio Sugiyama

Adult *amaurotis*. February, Aichi

Ashy Minivet

Pericrocotus divaricatus 19–20cm

Tokio Sugiyama

Adult ♂ *divaricatus*. May, Hegura-jima

Description Slender, aerial songbird with a long tail. Male has a black crown, white forehead and neck-sides, grey back, black wings, black tail with white outer feathers and white underparts. Female is greyer. Two races in Japan; *tegimae* (sometimes considered a separate species, Ryukyu Minivet *P. tegimae*) in southern Japan is darker, and has a black head with a narrow white forecrown and a black back.

Voice Loud, metallic trill *jirrrh-jirrrh*, calls frequently in flight.

Manabu Totsuka

Adult ♂ *tegimae*. April, Okinawa-Hontou

Range Breeds from the lower Amur Basin and Ussuriland south to Korea and Japan. Winters in south-east Asia.

Status in Japan Common summer visitor to broad-leaved deciduous forests on Honshu and Shikoku. Race *tegimae* is a common resident in evergreen forests from Yakushima southwards.

Tokio Sugiyama

Adult ♀ *divaricatus*. May, Hegura-jima

Goldcrest

Regulus regulus 9–10cm

Description Japan's smallest bird. Has a distinctive yellow stripe on crown; male's has orange in centre, which is often difficult to see. Also has an olive-green back, white eye-ring and lores, two wing-bars on dark wings and pale brown underparts.
Voice Song is a series of soft *trrrh* notes. Call is a thin *seeh-seeh*, similar to Coal Tit.

Range Breeds across temperate Eurasia from Europe east to Ussuriland and Sakhalin. Northern populations are migratory.
Status in Japan Common resident in coniferous forests from Honshu northwards. Breeds in subalpine areas in summer, moving to lower altitudes in winter. Often forms mixed flocks with Coal Tit in winter.

Tadao Shimba

Adult. January, Hokkaido

Bohemian Waxwing

Bombycilla garrulus 19–20cm

Description Silky, pink-tinged greyish-brown overall with a distinctive crest. Also has black wings with white bands on secondaries and primary coverts, yellow tips to primaries, waxy red tips to secondaries, and rufous undertail-coverts. Yellow tail–tip is distinctive.
Voice Soft, thin and sibilant *siree-ree*.
Similar species See Japanese Waxwing.

Range Breeds in northern Eurasia from Europe east to far north-eastern Russia and Kamchatka, south to Lake Baikal and lower Amur Basin; also in northern North America. Winters south to temperate regions.
Status in Japan Uncommon winter visitor to central Honshu and northwards. Numbers vary from year to year, and irruptions of large flocks occur. Occurs mainly in wooded areas in mountains, but flocks may visit city parks and gardens in search of berries.

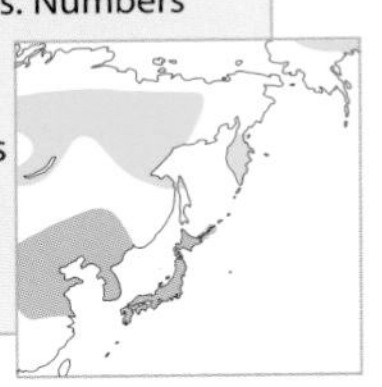

Tokio Sugiyama

Adult. February, Aichi

Tokio Sugiyama

Adult. March, Aichi

Japanese Waxwing

Bombycilla japonica 17–18cm

Tokio Sugiyama

Description Similar to Bohemian Waxwing, but lacks yellow tips to primaries and white wing-bands. Has a distinctive red-tipped tail, a yellowish belly, red-tipped greater coverts and reddish undertail-coverts.
Voice Soft, thin and high *see-see* or *sirr*.

Adult. April, Aichi

Tokio Sugiyama

Adult. April, Aichi

Range Breeds in south-eastern Russia, south to northern Ussuriland and in north-eastern China. Winters in eastern China, Korea and Japan.
Status in Japan Uncommon winter visitor to wooded areas; feeds on mistletoes and various berries. Gregarious. Most often seen in western Japan and flocks often visit parks and gardens in search of berries in early spring.

Tadao Shimba

Adult. March, Aichi

Brown Dipper

Cinclus pallasii 21–23cm

Description Chunky chocolate-brown bird of rivers and streams. Has a short tail, pointed dark bill and greyish legs. Immature is paler brown with white edges to wing-coverts.
Voice Song is a variety of ringing trills with grating notes, call is a loud and strong *dzeet-dzeet*.

Range Breeds from western Asia to eastern China and Taiwan, north to the Okhotsk coast and Kamchatka.
Status in Japan Common resident on rivers and streams from Kyushu northwards. Sedentary, spending all year on the same stretch of river. Flies low and direct over the water with rapid wingbeats. Often builds nest on a rock face behind a waterfall.

Tadao Shimba

Adult. April, Aichi

Winter Wren

Troglodytes troglodytes 10–11cm

Description Small, dark reddish-brown bird with short, cocked tail and pointed bill. Body finely barred dark-brown.
Voice Song is a long series of loud and musical ringing trills, delivered from an exposed perch; call is a dry *chek-chek*, similar to Japanese Bush Warbler, but harsher.

Range Widely distributed in northern Eurasia from Europe east to Kamchatka, south to Korea, Japan and Taiwan; also in North America and north Africa.
Status in Japan Common resident from Kyushu northwards. Breeds in mountain forests and prefers moist areas. Moves to lower altitudes in winter and inhabits dense shrubbery near streams.

Manabu Totsuka

Adult. May, Nagano

Alpine Accentor

Prunella collaris 17–19cm

Tokio Sugiyama

Adult. November, Shiga

Description Characteristic bird of high mountains. Has a dark-grey head, breast and hindneck; greyish-brown back with dark streaks, reddish-brown belly with white streaks, dark bill with yellow base to lower mandible, pinkish legs. Wings dark-brown with white tips to greater and median-coverts.

Voice Song is a variety of melodious trills with squeaky notes, call is a short *trru*.

Similar species Japanese Accentor is smaller, and has uniform dark–grey underparts.

Range Sparsely distributed in mountains in temperate Eurasia. In north-eastern Asia, breeds from the Amur Basin and Ussuriland south to northern Korea and central Japan.

Status in Japan Common in alpine areas in central and northern Honshu and moves to lower altitudes in winter. Prefers rocky areas; often approachable.

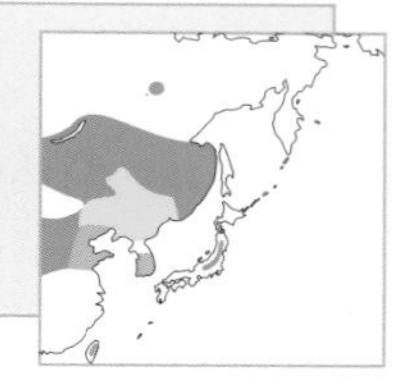

Siberian Accentor

Prunella montanella 14–16cm

Shigeo Ozawa

Adult. November, Hegura-jima

Description Has a dark crown and face with broad ochre-yellow supercilium, yellowish throat and grey nape. Brown back with dark streaks, buff underparts with chestnut flank streaks, thin dark bill and pinkish legs.

Voice Call is a loud and metallic *tirrre*, similar to Japanese Accentor.

Range Breeds in northern Siberia east to far north-east Russia, south to the Altai mountains and Amur Basin. Winters south to Mongolia, northern China and Korea.

Status in Japan Rare winter visitor to forest edges and reedbeds. Regular on Hegura-jima in November.

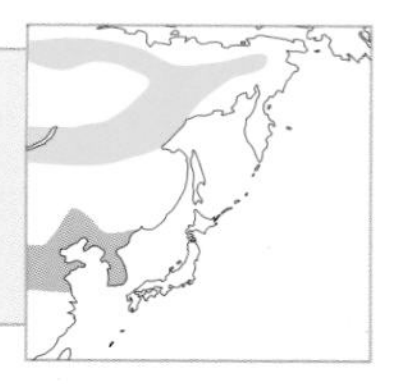

Japanese Accentor

Prunella rubida 14cm

Description A dark-coloured songbird that inhabits subalpine areas. Has a dark-grey face and underparts, greyish-brown head and brown back with dark streaks, thin dark bill and pinkish legs.
Voice Song is a melodious squeaky warble delivered from exposed places such as the top of a conifer.
Similar species Grey Bunting is slightly larger and greyer and has a stout two-toned bill.

Kazuyasu Kisaichi
Adult. June, Hokkaido

Range Breeds in northern Japan and the southern Kurile Islands. Winters south to southern Japan.
Status in Japan Breeds in subalpine areas in Shikoku, Honshu and Hokkaido. Moves to lower altitudes in winter and prefers dense shrubbery, often near streams.

Tadao Shimba
Adult. December, Aichi

White-throated Rock Thrush

Monticola gularis 18.5cm

Description Male has a blue crown, chestnut face with white throat and dark ear-coverts, dark back and chestnut underparts, blue spot on wing-coverts and a small white wing patch. Female is pale greyish-brown above with dark scaly marks, a white throat and whitish underparts heavily barred dark-brown.
Similar species Female Blue Rock Thrush is larger and darker. Female Rufous-tailed Rock Thrush has a rusty-coloured tail.

Teruaki Ishi
Adult ♂. May, Hegura-jima

Range Breeds from south-eastern Transbaikalia east to Ussuriland, south to north-eastern China and Korea. Winters in southern China and south-eastern Asia.
Status in Japan Rare passage migrant on coasts of the Sea of Japan.

Blue Rock Thrush

Monticola solitarius 25.5cm

sighted 1/15/29 in Fiddler's Green

Tadao Shimba

Adult ♂ breeding. May, Hegura-jima

Description Male is unmistakable, with its blue head, breast and back, dark-brown wings and chestnut belly. Female has dark-brown upperparts and buff-coloured underparts with heavy dark-brown spots. Male of vagrant race *pandoo* has blue underparts.

Voice Song is a loud melodious warble.

Similar species Female Rufous-tailed Rock Thrush *M. saxatilis* (recorded once in Japan) is similar to female, but smaller and paler with a rusty-coloured tail and shorter bill.

Shuichi Tsuno

Adult ♂ breeding. April, Wakayama

Range Widespread across southern Eurasia from southern Europe east to south-east Asia, north to Ussuriland; also in north Africa.

Status in Japan Common resident in rocky coastal areas and also common around human habitations near coasts. Northern populations move south in winter.

Tadao Shimba

Immature ♂. August, Chichi-jima

Tadao Shimba

Juvenile. August, Chichi-jima

Siberian Thrush

Zoothera sibirica 23.5cm

Description Male is unmistakable: sooty-black with distinctive white supercilium and white bars on lower belly and undertail-coverts. First year male has buff-coloured edges to wings and white spots on undertail-coverts. Female has olive-brown back, pale supercilium, mottled ear-coverts, pale moustachial stripe, white throat and buff-coloured spots on underparts. In flight shows conspicuous broad white-bordered black band on underwing.

Voice Song is distinctive disyllabic, loud, short fluty whistle and quiet chirping. Call is a harsh *seeeh*.

Similar species Female Japanese Thrush is similar to female but has heavy dark spots on underparts and rusty flank patches.

Mitsuo Imai

Adult ♂. May, Hegura-jima

Range Breeds from southern Siberia east to Sakhalin and the Okhotsk coast, south to north-eastern China and Korea. Winters in southern China and south-east Asia.

Status in Japan Uncommon summer visitor to mixed forests in mountains and highlands from central Honshu northwards; rare passage migrant elsewhere.

Tadao Shimba

Adult ♀. May, Hegura-jima

Amami Thrush

Zoothera major 30cm

Kazuyasu Kisaichi

Adult. April, Amami-Oshima

Description Similar to White's Thrush, but darker with large thicker bill and exposed bare pink skin behind the eye.
Voice Song is a loud, short melodious warble and different from that of White's Thrush. Sings mostly at dawn.
Similar species Identification needs care in winter when White's Thrush visits the Amami Islands.

Range Endemic to Amami Islands in Japan.
Status in Japan Rare resident in dense sub-tropical forests on Amami Islands. Normally shy, skulks on the ground in dense cover. Occasionally revealed roosting in trees by spotlighting.

Himaru Iozawa

Adult. September, Amami-Oshima

White's Thrush

Zoothera aurea 30cm

Description Unmistakable large thrush; sexes alike. Has yellowish-brown head and back, and whitish underparts with heavy, black, crescent-shaped markings. In flight shows a conspicuous broad white-framed black band on underwings and also white
tail corners.
Voice Song is plaintive, whistling note, lasting about a second *hyyyyy*.
Similar species Amami Thrush is darker with a thicker bill and a different song.

Range Breeds from southern Siberia east to Ussuriland, south to north-eastern China and Korea. Winters in southern China and south-east Asia.
Status in Japan Uncommon resident in wooded areas. Moves south in winter and often appears in city parks and gardens. Summer visitor to Hokkaido and winter visitor to the Ryukyu Islands. Normally shy and skulks on the ground in dense cover.

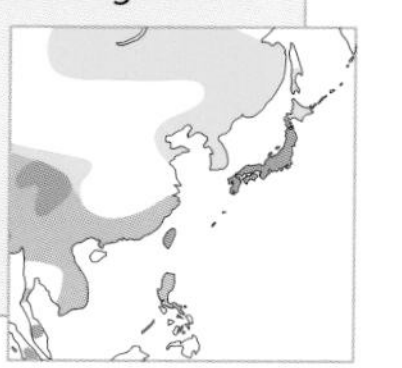

Manabu Totsuka

Adult. February, Aichi

Tokio Sugiyama

Adult. May, Hegura-jima

Kazuyasu Kisaichi

Adult. February, Chiba

Grey-cheeked Thrush

Catharus minimus 16-17cm

Tuomo Jaakkonen

Adult. June, Chukotka, Russia

Description Small thrush with grey cheeks and bold dark spots on breast. Uniform olive-brown above with whitish underparts and grey flanks.
Voice Call is a harsh *weer*. Song is a rolling series of descending nasal notes delivered from high in tree.

Range Mainly Nearctic; this species is the only *Catharus* thrush to have extended its range into north-east Asia. This population breeds from Chukotka west to the Kolyma; the birds migrate south through North America.
Status in Japan Vagrant. One on Hegura-jima in September 2004.

Song Thrush

Turdus philomelos 22cm

Teruaki Ishii

First-winter. November, Kanagawa

Description An olive-brown thrush with pale eye-ring and white underparts with heavy arrow-shaped dark spots. In flight shows pale reddish-buff underwing–coverts.
Voice Short and metallic *zit* or *tik*, similar to that of some buntings.
Similar species Mistle Thrush is larger, has greyish fringes to wings and upperparts, and a long white-tipped tail.

Range Widespread in Europe, east through Siberia to about Lake Baikal. Northern populations are migratory and winter south to northern Africa and south-western Asia.
Status in Japan Vagrant. One record in 1987 in Kanagawa.

Fieldfare

Turdus pilaris 25.5cm

Description Unmistakable large thrush with grey head, ear-coverts and hindneck; reddish-brown back, yellow bill, dark legs, grey rump and uppertail-coverts and whitish underparts, heavily barred on breast and flanks. Sexes alike.
Voice Call is a chattering *chack-chack-chack*.

Tokio Sugiyama

Adult. February, Kanagawa

Range Breeds from central Europe and Scandinavia east through Siberia to the Lena and Lake Baikal. Winters further south in Europe, and in central and south-western Asia.
Status in Japan Vagrant with a few records from Honshu and Hokkaido. Found in cultivated fields and open forests.

Mistle Thrush

Turdus viscivorus 27cm

Description Large thrush with long tail and white outer tail feathers. Pale greyish-brown above; whitish below with heavy dark spots.
Voice Call is a dry rattling *trrrrrh*.
Similar species Song Thrush is browner, smaller and has a shorter tail.

Range Widespread in Europe, east through Siberia to about Lake Baikal. Northern populations are migratory and winter south to northern Africa and south-western Asia.
Status in Japan Vagrant. First recorded in February 1984 in Aichi and a few other sightings since.

Akira Yamamoto

Probable first-winter. February, Aichi

Grey-backed Thrush

Turdus hortulorum 23cm

Mitsuo Imai

Adult ♂. February, Kanagawa

Tadao Shimba

Adult ♀. May, Hegura-jima

Description Male has greyish-blue upperparts and breast, white belly, rufous flanks, yellowish undertail-coverts, a yellow bill and pinkish yellow legs. Female has greyish-brown upperparts, a white throat with heavy dark streaks, and rufous flanks.

Voice Song consists of loud and melodious whistles. Call is a harsh *zeeet*.

Similar species Female Japanese Thrush has a browner back and more extensive dark brown spots on underparts.

Range Breeds from far south-eastern Russia to north-eastern China. Winters in south China and northern south-east Asia.

Status in Japan Rare passage migrant to wooded areas; mainly seen in spring on the coasts of the Sea of Japan.

Tokio Sugiyama

First-summer ♂. May, Hegura-jima

Japanese Thrush

Turdus cardis 21.5cm

Description Male has a black head, breast and upperparts; yellow bill and eye-ring, white belly with arrow-shaped dark spots. Female has dark-brown back, white throat and white underparts with arrow-shaped dark spots and rusty flank patches.
Voice Song is a loud and melodious whistle, *tjew tuip tuip tuip tuip,* repeated with many variations; delivered from the canopy of broad-leaved deciduous trees where the birds are often hard to see. Call is a harsh *zeeet.*
Similar species Female is similar to female Grey-backed Thrush but is browner and more heavily spotted below.

Range Breeds in Japan and central China and winters in southern China and northern part of south-eastern Asia.
Status in Japan Common summer visitor to broad-leaved deciduous forests from Kyushu northwards. Some winter in southern Japan. Feeds mainly on worms, but also feeds on berries in autumn.

Tokio Sugiyama

Adult ♂. May, Hegura-jima

Tokio Sugiyama

Adult ♀. May, Hegura-jima

Tadao Shimba

First-winter ♂. February, Aichi

Blackbird

Turdus merula 25–28cm

Kenji Takehara

Adult ♂. March, Yonaguni-jima

Manabu Totsuka

Adult ♀. January, Okinawa-Hontou

Description Male is all black with a yellow bill and eye-ring. Female has a dark brown body with a slightly paler throat and breast, pale-brown bill and indistinct eye-ring.
Voice Song is a variety of fluty whistles; call is a harsh *chak-chak*.
Similar species Female Siberian Thrush is similar to female, but has pale supercilium, black bill, yellowish-brown spotted underparts and yellowish legs.

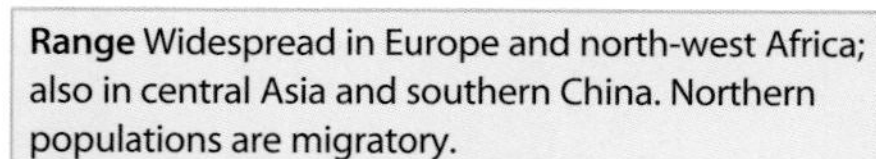

Range Widespread in Europe and north-west Africa; also in central Asia and southern China. Northern populations are migratory.
Status in Japan Rare visitor to cultivated and grassy fields and forest edges. Most records are from Ryukyu Islands; regularly seen on Yonaguni-jima in spring. Often feeds around open forest edges.

Redwing

Turdus iliacus 20–21cm

Mitsuo Imai

Adult. January, Chiba

Description Dark-brown above with a conspicuous supercilium, dark malar stripe and dark bill with yellow base to lower mandible. White below with heavy, dark streaks and rusty flank patches. In flight shows conspicuous reddish underwing-coverts.
Voice Call is a thin, harsh *seeeh*; also harsh *chuk-chuk*.
Similar species Both female Japanese Thrush and female Grey-backed Thrush lack white supercilium; *naumanni* Dusky Thrush is larger and has a rusty breast.

Range Breeds from northern Europe east through Siberia to the Kolyma River and Lake Baikal. Winters south to central and southern Europe, and western and central Asia.
Status in Japan Vagrant. Several records from open forests and cultivated fields.

Eyebrowed Thrush

Turdus obscurus 21.5cm

Description Male has dark bluish-grey head and throat with prominent white supercilium and white line across cheek; dark bill with yellow lower mandible, olive-brown back and tail, rufous breast and flanks, white belly and white undertail-coverts. Female is paler, has white supercilium, chin stripe and moustachial stripe.
Voice Call is a harsh thin *seeeh*.
Similar species Brown-headed Thrush lacks prominent white supercilium.

Tokio Sugiyama

First-summer ♂. April, Aichi

Range Breeds from central Siberia east to the Okhotsk coast and Kamchatka Peninsula, south to Transbaikalia and Amur Basin. Winters in southern China, Taiwan and south-east Asia.
Status in Japan Uncommon passage migrant to woodlands; a few winter in the south. Arrives as early as September, ahead of other wintering thrushes.

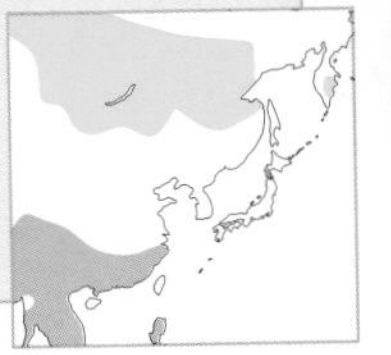

Kazuyasu Kisaichi

First-summer ♀. May, Hegura-jima

Tokio Sugiyama

Adult ♂. April, Aichi

Pale Thrush

Turdus pallidus 23–25cm

Tokio Sugiyama

Adult ♂. January, Kyoto

Description Male has dark-greyish head and throat, olive-brown back and dark tail with white corners. Also buffy-grey on breast and flanks with white belly and undertail-coverts. Female has olive-brown head and back, pale supercilium, white throat with brown moustachial stripe and paler underparts.
Voice Song is a loud melodious whistle and call is a harsh *zeeeh* or loud and explosive *kwak-kwak-kwak* when alarmed.
Similar species Brown-headed Thrush has rufous flanks and no white on tail.

Tadao Shimba

Adult ♀. February, Aichi

Range Breeds from the Amur basin east to Ussuriland, south to north-eastern China and Korea. Winters in southern Korea, Japan, Taiwan and south-eastern China.
Status in Japan Common winter visitor to wooded areas from central Honshu southwards. Common in city parks and wooded areas. A few breed in Hiroshima.

Yoshihito Goto

Adult ♂. February, Amami-Oshima

Brown-headed Thrush

Turdus chrysolaus 23.5cm

Description Male has dark-brown head, olive-brown back and tail, rufous breast and flanks, white belly and undertail-coverts, a dark bill with yellow lower mandible and yellowish legs. Female is duller than male and has a white throat with brown streaks. Two races in Japan; *orii* is slightly larger than the nominate *chrysolaus*, and has a darker head.
Voice Melodious warbling song, often beginning with a beautiful trill, and then progressing into whistling series of notes. Call is a harsh *zeeeh*.
Similar species Izu Islands Thrush has reddish-brown back; male has prominent black hood.

Tadao Shimba

Adult ♂ *chrysolaus*, June, Nagano

Range Breeds on Sakhalin, Kurile Islands, and northern and central Japan. Winters south to Taiwan and eastern China.
Status in Japan Common summer visitor to deciduous forests from central Honshu northwards. Winters from central Honshu south. Two races; orii breeds on Kuril Islands and is an uncommon winter visitor to Japan. Nominate *chrysolaus* breeds in Hokkaido and northern Honshu.

Tadao Shimba

Adult ♀. May, Hegura-jima

Kazuyasu Kisaichi

Adult ♂ *orii*, January, Ibaragi

Izu Islands Thrush

Turdus celaenops 23cm

Kazuyasu Kisaichi

Adult ♂. April, Hachijo-jima

Kazuyasu Kisaichi

Adult ♀. April, Hachijo-jima

Description Male has a black hood and bright yellow bill and eye-ring; reddish-brown back, dark wings and tail, rufous belly and rufous flanks. Female has dark brown head, yellowish bill and eye-ring, and reddish-brown back; white throat with brown streaks, white belly and undertail-coverts with brown spots.
Voice Song is a harsh ratchet-like *ki ree ree ree ree*. Call is a harsh *zeeeh*.
Similar species Brown-headed Thrush has olive-brown back, darker bill and white undertail-coverts that lack brown spots.

Range Endemic to southern Japan.
Status in Japan Common resident in evergreen forests on Izu Islands, rare resident on Tokara Islands and Yakushima. In winter, some visit Pacific coastal areas of central Honshu.

Akira Yamamoto

Adult ♂. April, Miyake-jima

Dark-throated Thrush

Turdus ruficollis 23–24cm

Description Two races (sometimes treated as separate species: Black-throated Thrush *T. atrogularis* and Red-throated Thrush *T. ruficollis*), both of which occur in Japan. Male *atrogularis* has slate-grey back, black throat and breast and white belly, and dark bill with yellow lower mandible. Female and first–year male have breast heavily streaked. Male *ruficollis* has rufous supercilium, throat and breast; female has pale supercilium, black malar stripe and dark streaks on breast, also rufous patches on breast and belly.
Similar species Dusky Thrush of race *naumanni* is browner with rusty spots on breast, undertail-coverts and tail.

Range *atrogularis* breeds in western Siberia and central Asia and winters from Iran to Indian subcontinent; *ruficollis* breeds in the Altai and Transbaikalia and winters from north-east India to western China.
Status in Japan Rare visitor to open forests and fields.

Yoshio Goto

Adult ♂ *ruficollis*. November, Hegura-jima

Kazuyasu Kisaichi

First-winter *ruficollis*. April, Osaka

Teruaki Ishii

Dusky Thrush

Turdus naumanni 24cm

Tadao Shimba

Adult ♂ *eunomus*. February, Aichi

Description Two races (sometimes split as Dusky Thrush *T. eunomus* and Naumann's Thrush *T. naumanni*); both occur in Japan. Race *eunomus* has dark greyish-brown head, back, chestnut wings and brown tail. Has a distinctive cream-coloured supercilium and throat, dark bill with yellow at base, white underparts with dark breast-band and dark arrow-shaped spots on flanks; female is paler with brown wings. Race *naumanni* has reddish-brown supercilium, breast, belly, tail and undertail-coverts.
Voice Call is a harsh and distinctive *kwat-kwat*, or loud harsh *trrrrh*.
Similar species Female *ruficollis* Dark-throated Thrush is greyer, and has greyish-brown uppertail-coverts and white undertail-coverts.

Tadao Shimba

Adult ♀ *eunomus*. May, Hegura-jima

Range Race *eunomus* breeds in northern Siberia from Yenisey east to Kamchatka; *naumanni* breeds in south and central Siberia. Both races winter in eastern China, Taiwan, Korea and Japan.
Status in Japan Race *eunomus* is abundant winter visitor to open forests, cultivated fields and wooded urban areas; *naumanni* is a rare visitor, but regularly seen during spring and autumn migration on coasts of the Sea of Japan.

Kazuyasu Kisaichi

Adult ♂ *naumanni*. February, Tokyo

Norio Yamagata

Adult ♀ *naumanni*. May, Hegura-jima

Tadao Shimba

Zitting Cisticola

Cisticola juncidis 12.5–13.5cm

Description Small, active bird of reedbeds and rank grassland. Yellowish-brown above, heavily streaked blackish on crown, mantle, back and uppertail-coverts; pale supercilium and underparts; short, rounded tail with white tip and dark subterminal band. In non-breeding plumage has white streaks on crown and back. Male performs song flight in breeding season.
Voice Song is a metallic and quickly repeated *zip-zip* when flies up from the grass and an occasional harsh *chet-chet* when descending.
Similar species Japanese Swamp Warbler is larger with heavily streaked reddish-brown back and longer tail.

Range Widespread in southern Europe, Africa, Indian subcontinent, south-east Asia and Australia. Northern populations are migratory. Uncommon summer visitor to Korea.
Status in Japan Common resident in reedbeds and grasslands from central Honshu southwards. Moves south in winter.

Tokio Sugiyama

Adult breeding. June, Aichi

Tadao Shimba

Adult non-breeding. September, Okinawa-Hontou

Akira Yamamoto

Adult breeding. June, Aichi

Asian Stubtail

Urosphena squameiceps 10.5cm

Tokio Sugiyama

Adult. July, Aichi

Description Secretive, small and extremely short-tailed warbler. Dark-brown above with dark eye-stripe and long buff-coloured supercilium with dark stripe above. Whitish underparts with brown wash, dark bill and pinkish legs.
Voice Song is high, insect-like and quickly repeated *tsit-tsit-tsit-tsit* becoming louder.

Shigeo Ozawa

Probable juvenile. October, Tokyo

Range Breeds from Ussuriland, Sakhalin south to north-eastern China and Korea. Winters in southern China, northern parts of south-east Asia and Taiwan.
Status in Japan Common summer visitor to deciduous and mixed forests from Kyushu northwards. Prefers moist areas. Small numbers winter on the Ryukyu Islands.

Manchurian Bush Warbler

Cettia canturians 16cm

Himaru Iozawa

Adult. March, Yonaguni-jima

Description Similar to Japanese Bush Warbler but has reddish-brown crown and back, more prominent supercilium, greyish cheeks, larger brown bill and longer legs.

Range Breeds in north-eastern China and Korea, winters in southern China and south-east Asia.
Status in Japan Rare winter visitor to Yaeyama Islands. A few sightings on offshore islands in Sea of Japan. Occurs in wooded areas with thickets.

Japanese Bush Warbler

Cettia diphone 14–16cm

Description Olive-brown back, long tail, pale supercilium, brown under-parts, brown bill and pinkish legs. **Voice** Song a vibrating low whistle *hooooo* followed by loud *ho-tjee-o*, often followed by long series of *te-chew te-chew*. Call is a harsh *chet-chet*. The distinctive song is very familiar to local people and is associated with the coming of spring. **Similar species** Manchurian Bush Warbler is slightly larger, has longer bill and legs and reddish-brown back.

Tadao Shimba

Adult. January, Okinawa-Hontou

Range Sakhalin south to Japan. **Status in Japan** Common resident. Breeds from foothills to montane forests with dense undergrowth. In winter, moves to warmer areas and often appears in city parks and gardens. Race *diphone* on Ogasawara Islands has a longer bill.

Tokio Sugiyama

Adult *cantans*. May, Hegura-jima

Yoshihito Goto

Adult *diphone*. August, Haha-jima

Tokio Sugiyama

Adult *cantans*. May, Hegura-jima

Spotted Bush Warbler

Bradypterus thoracicus 13cm

David Fisher

Adult. Qinghai, China

Description Olive-brown warbler with short, rounded wings. Pale supercilium, dark eye-stripe and mottled ear-coverts. White throat, greyish wash on breast with blackish spots, whitish underparts with buff flanks and undertail-coverts.
Similar species Separated from other warblers by blackish breast spots.
Note Northern population is sometimes separated as Baikal Bush Warbler *B. davidi*.

Range Southern Siberia east to south-east Russia and north China, south to northern Korea, also Himalayas and central China. Moves south in winter. Grassy fields and open forests.
Status in Japan No records

Chinese Bush Warbler

Bradypterus tacsanowskius 13cm

Jens S. Hansen

Adult. Yang He River, Bedaihe, China

Description Small warbler, olive-brown above with whitish supercilium and lores and short, rounded wings. Whitish throat and belly, buff-coloured breast and flanks, long broad tail and pinkish legs.
Voice Song is an insect-like rasping chirruping.
Similar species Spotted Bush Warbler has spotted breast and white chevron markings on undertail-coverts.

Range Breeds in eastern Siberia to Ussuriland and north-eastern China, south to central China. Winters south to southern Asia. Nests in dense cover in high mountains. Secretive and hard to observe.
Status in Japan No records.

Lesser Whitethroat

Sylvia curruca 13cm

Description Grey head, greyish-brown back with dark-brown ear-coverts and broken white eye-ring. White throat and white underparts with pale greyish-brown flanks, dark bill and dark legs.
Similar species Separated from Common Whitethroat *S. communis* by slightly smaller size and darker bill, legs, ear-coverts and upperparts.

Yoshio Osawa

Adult. February, Yamagata

Range Breeds across Eurasia from Europe east to the Lena River in Siberia. Winters from tropical Africa east to the Indian subcontinent.
Status in Japan Vagrant, with a few records from Hokkaido and Honshu. Found in wooded areas with thickets.

Chinese Hill Warbler

Rhopophilus pekinensis 18cm

Description Large warbler with short, rounded wings and long, graduated tail. Has a distinctive black lores and moustachial stripe, pale supercilium and yellowish, slightly downcurved bill. Heavily black-streaked crown and back; white underparts streaked rufous on breast-sides and flanks; white tips to rectrices.

Ran Schols

Adult. May, Hebei, China

Range Breeds in western and northern China east to northern Korea. Inhabits open areas of stony, bushy hills; local and uncommon. Very active and often observed in small flocks, busily foraging in lower bushes.
Status in Japan No records.

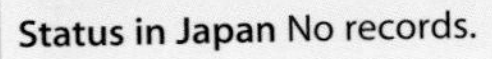

Lanceolated Warbler

Locustella lanceolata 12cm

Manabu Totsuka

Adult. June, Hokkaido

Description Small brown grasshopper warbler with heavily dark-streaked head, back and rump. Pale supercilium, white underparts with finely streaked throat and breast, brownish flanks with dark streaks.
Voice Song is a cricket like thin and endless trill *trrrrrrrrrh.*
Similar species Pallas's Grasshopper Warbler is slightly larger with reddish-brown back and white tips to rectrices.

Kazuyasu Kisaichi

Adult. July, Hokkaido

Range Breeds across Eurasia from eastern Europe east through Siberia and Mongolia to north-east China and Sakhalin. Winters in southern Asia. Uncommon passage migrant in Korea.
Status in Japan Common summer visitor to grassy fields and freshwater marshes in Hokkaido; rare passage migrant elsewhere. A few breeding records from central Honshu.

Teruaki Ishii

Adult. June, Hokkaido

Middendorff's Grasshopper Warbler

Locustella ochotensis 15.5cm

Description Olive-brown above with indistinct dark streaks. Pale supercilium and white underparts with brown wash on breast and flanks, white tips to rectrices.
Voice Song is a loud, high-pitched warble *chew-chew-chew* delivered from an exposed perch on the top of a thicket or in short song flight.
Similar species Styan's Grasshopper Warbler has greyish-brown back, longer bill and legs; also has longer tail with smaller white tips to rectrices.

Range Breeds from Kamchatka Peninsula south to the Okhotsk coast and northern Japan. Winters in Indonesia and Philippines.
Status in Japan Common summer visitor to Hokkaido; uncommon passage migrant elsewhere. Breeds on coastal grasslands and in freshwater marshes with thickets.

Manabu Totsuka

Adult. June, Hokkaido

Tadao Shimba

Adult. July, Hokkaido

Yoshio Goto

Juvenile. September, Hegura-jima

Styan's Grasshopper Warbler

Locustella pleskei 17cm

Tadao Shimba

Adult. July, Wakayama

Description Greyish-brown above with pale supercilium, white throat and underparts, brown flanks, long dark bill, pinkish legs and a long tail with small white tips to rectrices.
Voice Song is similar to Middendorff's Warbler.
Similar species Middendorff's Grasshopper Warbler is browner with shorter bill, legs and tail.

Tadao Shimba

Adult. July, Wakayama

Range Breeds locally in Ussuriland, coastal Korea and central Japan. Winters in south-eastern China and Vietnam.
Status in Japan Locally common summer visitor to coastal thickets. Breeds on the Izu Islands and in coastal areas of Mie, Wakayama and southern Kyushu.

Tadao Shimba

Adult. July, Wakayama

Gray's Grasshopper Warbler

Locustella fasciolata 18cm

Description Large warbler, dark olive-brown above with rusty rump and tail. White supercilium, whitish underparts and olive-brown flanks.
Voice Song is loud and distinctive *tyut tjit tji tju-tji hue.* Song is delivered from dense cover; sings by night and day.
Similar species Separated from other grasshopper warblers by larger size and distinctive song.

Range Breeds in southern Siberia east to Sakhalin and Kuril Islands; south to north-eastern China and Korea. Winters in south-east Asia.
Status in Japan Common summer visitor to Hokkaido; rare passage migrant elsewhere. Inhabits open forests with dense cover; hard to observe.

Kazuyasu Kisaichi

Adult. July, Hokkaido

Kazuyasu Kisaichi

Adult. July, Hokkaido

Kazuyasu Kisaichi

Adult. June, Chiba

Pallas's Grasshopper Warbler

Locustella certhiola 12–13.5cm

Björn Anderson

Adult. May, Bayan Nuur, Mongolia

Description Dark olive-brown above with a heavily-streaked crown and back. Pale supercilium; short, rounded wings and reddish-brown rump and uppertail-coverts. Whitish underparts with buff-coloured breast, flanks and undertail-coverts; broad tail with dark subterminal band and white tips to rectrices.
Voice Call is a thin *tik-tik.*
Similar species Lanceolated Warbler has finely streaked upperparts and streaked breast and flanks, a thinner bill and no white on tail.

Yoshiki Watabe

Adult. May, Hegura-jima

Range Breeds from northern Siberia east to Okhotsk coast and Ussuriland, south to Mongolia and northern China. Winters in southern Asia.
Status in Japan Rare passage migrant to grassy fields and undergrowth in coastal areas of Sea of Japan. A few regularly winter on Yonaguni-jima.

Pavel Parkhaev

Adult. July, Ust'-Kan, Altai, Russia.

Japanese Marsh Warbler

Locustella pryeri 13–14cm

Other name Marsh Grassbird
Description Has a reddish-brown head and back with dark streaks and a long, graduated tail with pointed feather tips. Dark bill with pale lower mandible, pale underparts and pinkish legs. Male performs song flight in breeding season.
Voice Song is a series of chirrups.
Similar species Pallas's Grasshopper Warbler is darker with white-tipped rectrices.

Range Breeds in Ussuriland, north-eastern China and central Japan. Winters in eastern China and central Honshu.
Status in Japan Locally common breeding resident in Chiba and Ibaragi; also a summer visitor to northern Honshu. Winters south to southern Honshu. Reedbeds and grassy fields with thickets.

Kazuyasu Kisaichi

Adult. June, Chiba

Kazuyasu Kisaichi

Adult. July, Chiba

Norio Yamagata

Adult. June, Aomori.

Black-browed Reed Warbler

Acrocephalus bistrigiceps 12.5–13.5cm

Tadao Shimba

Adult. June, Hokkaido

Description Olive-brown above, white supercilium, dark eye-stripe and prominent dark brown lateral crown-stripe; whitish underparts with buff-coloured flanks. Juvenile is similar to adult, but underparts more buff-coloured.
Voice Song is loud, a variety of metallic warbles and chirruping.
Similar species Other reed warblers lack prominent black crown-stripe.

Tadao Shimba

Adult. June, Hokkaido

Range Breeds in south-eastern Siberia and Sakhalin south to north-eastern China and Korea. Winters in southern China and northern south-east Asia.
Status in Japan Common summer visitor to reedbeds and grassy fields from central Honshu northwards. Breeds in montane meadows in Kyushu.

Hyun-tae Kim

Juvenile. September, Sosan, South Korea

Paddyfield Warbler

Acrocephalus agricola 13cm

Description Olive-brown above; white supercilium bordered above by dark brow-line, dark eye-stripe and indistinct eye-ring. Narrow supercilium is prominent above and behind eye. White below with buff-coloured tinge to sides of breast and flanks; short wings and long tail.
Voice Song is a rapid series of melodious phrases without harsh notes. Mimetic.
Similar species Black-browed Reed Warbler has more prominent black brow above supercilium, and a longer tail. See also Manchurian Reed Warbler.

Nobuo Dobashi

Adult. May, Hegura-jima

Range Breeds from eastern Europe to Mongolia and western China. Winters mainly in the Indian subcontinent. Inhabits reedbeds and grassy swamps.
Status in Japan Vagrant; one record on Hegura-jima in May 1992.

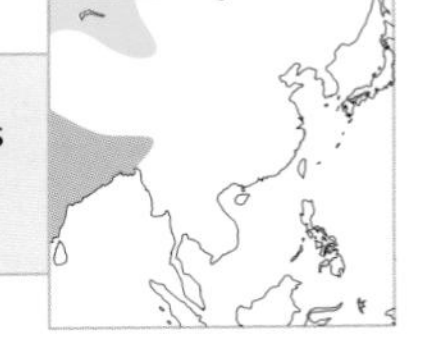

Manchurian Reed Warbler

Acrocephalus tangorum 13cm

Description Similar to Paddyfield Warbler, but has a longer, stouter bill with paler lower mandible. Prominent white supercilium is wider above and in front of eye and has a darker brow-line. Formerly treated as a race of Paddyfield Warbler.
Voice Similar to Paddyfield Warbler.
Similar species See Paddyfield Warbler and Black-browed Reed Warbler.

Range Breeds in Ussuriland and north-eastern China, winters in south-east Asia.
Status in Japan No records.

Leif Gabrielsen

Adult. May, Hebei, China

Blyth's Reed Warbler

Acrocephalus dumetorum 13cm

Tomi Muukkonen

Adult. May, Haapsalu, Finland

Description Greyish-brown head and back with pale lores and faint white eye-ring; whitish underparts.
Similar species Difficult to separate from Marsh Warbler *A. palustris,* (unrecorded in Japan) except by song and wing formula; Blyth's Reed Warbler has shorter outer primaries (P7 to P9).

Range Breeds across Eurasia from Finland east to central Siberia and north-western Mongolia. Winters in southern Asia.
Status in Japan Vagrant. Captured a few times at ringing stations in Hokkaido and Kume-jima.

Thick-billed Warbler

Acrocephalus aedon 20cm

Norio Yamagata

First-winter. February, Shizuoka

Kazuyasu Kisaichi

First-winter. March, Shizuoka

Description Large reed warbler with short, rounded wings. Rusty-brown above; face plain brown without supercilium; stout pale bill and whitish underparts with buff-coloured flanks.
Similar species Oriental Reed Warbler has browner back, longer bill, shorter tail and a pale supercilium.

Range Breeds in southern Siberia, northern Mongolia and north-eastern China. Winters in southern Asia.
Status in Japan Vagrant. Several records from Honshu and Kyushu. Frequents reedbeds and thickets.

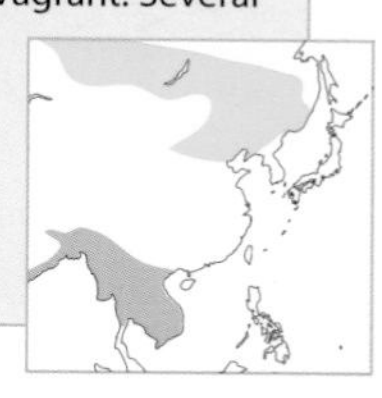

Oriental Reed Warbler

Arcocephalus orientalis 18–19cm

Description Large reed warbler, olive-brown above with pale supercilium. Dark bill with orange lower mandible, underparts whitish.
Voice Loud and variable, raucous, unmusical croaking song *gruw-gruw-eek kru-kih kru-kih kjew-kjew-kjew*. Song is delivered from the top of bushes or reeds.
Similar species Thick-billed Warbler has shorter, stouter bill, a rusty-brown back, a longer tail and lacks a supercilium. Black-browed Reed Warbler is smaller and has prominent blackish crown-stripes.

Tadao Shimba

Adult. May, Aichi

Range Breeds from south-eastern Siberia south to eastern China and Korea. Winters in southern Asia.
Status in Japan
Common summer visitor to grassy swamps, reedbeds and river banks from Kyushu northwards.

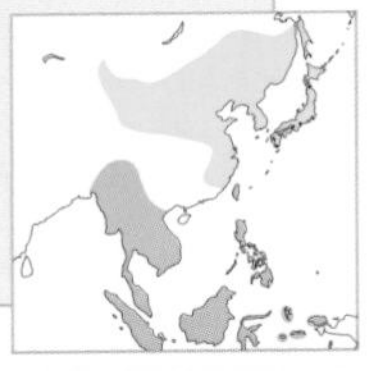

Tadao Shimba

Adult. June, Aichi

Tadao Shimba

Adult. May, Aichi

Manabu Totsuka

Adult. May, Aichi

Willow Warbler

Phylloscopus trochilus 11.5cm

Yoshihito Goto

First-winter. September, Hegura-jima

Description Greyish olive-brown above with a long primary projection and yellowish supercilium. Whitish below with yellow wash on throat and breast; brownish legs.
Voice Call is a soft *hu eeet.*
Similar species Siberian Chiffchaff is greyer with shorter primaries and dark legs.

Range Breeds across Eurasia from Europe east to far north-eastern Russia. Winters mainly in tropical and southern Africa.
Status in Japan Very rare autumn visitor, recorded mainly in coastal areas of Sea of Japan. Occurs in wooded areas.

Pale-legged Leaf Warbler

Phylloscopus tenellipes 11–12cm

Hyun-tae Kim

Adult. April, Oeyeon Island, South Korea

Description Dark greyish crown, greenish-brown above with two pale buffish wing-bars. Also a white supercilium, dark eye-stripe, mottled ear-coverts and white underparts; pinkish legs.
Voice Song is a cricket–like chirruping.
Similar species Separated from Sakhalin Leaf Warbler mainly by song and breeding range.

Range Breeds from Ussuriland south to northern Korea. Winters in south-east Asia.
Status in Japan No records.

Siberian Chiffchaff

Phylloscopus tristis 11cm

Description Greyish-brown above with a dark eye-stripe and pale supercilium. Whitish below. Bill and legs black.
Voice Call is a soft, monotonous *heet.*
Similar species Willow Warbler is browner with a longer primary projection, a yellow wash on throat and breast and brown legs.

Mitsuo Imai

First-winter. November, Hegura-jima

Range Breeds across Asia from the Urals east to the Kolyma River in Siberia. Winters in India.
Status in Japan Very rare visitor, recorded mainly in coastal areas of the Sea of Japan. Occurs in wooded areas.

Wood Warbler

Phylloscopus sibilatrix 12.5cm

Description Yellowish-green warbler with a bright yellow face, throat and breast. Has long, pointed wings with yellowish-green fringes to dark primaries and a white belly. Bill and legs yellowish-brown.
Voice Call is a short and metallic *zip*.
Similar species Arctic Warbler has a prominent pale supercilium and a whitish throat and breast.

Range Breeds from Western Europe east to western Siberia and winters mainly in tropical Africa.
Status in Japan Vagrant. Recorded in Hokkaido and Hegura-jima. Occurs in wooded areas.

Yoshihito Goto

First-winter. October, Hegura-jima

Dusky Warbler

Phylloscopus fuscatus 11–12cm

Tadao Shimba

Adult. May, Hegura-jima

Description Small greyish-brown warbler with light-brown underparts. The prominent supercilium is white in front of the eye, becoming wider and buffish behind the eye. Also a dark-brown eye-stripe, faint eye-ring and a dark bill with yellow base of lower mandible.

Voice Call is a short, harsh *chok-chok*.

Similar species Radde's Warbler has a larger head, browner upperparts and a shorter and thicker bill; also supercilium is wider and buff-coloured in front of eye. Japanese Bush Warbler is larger, has browner back and less marked supercilium.

Tadao Shimba

Adult. May, Hegura-jima

Range Breeds from Siberia east to the Kamchatka Peninsula, south to Sakhalin, Ussuriland and north-eastern China. Winters in southern China and south-east Asia.

Status in Japan Rare passage and winter visitor to wooded areas, thickets and reed beds. Regularly observed on offshore islands in Sea of Japan in spring. A few winter on Yaeyama Islands. Skulks in undergrowth and difficult to observe.

Kazuyasu Kisaichi

Adult. March, Chiba

Radde's Warbler

Phylloscopus schwarzi 12–13cm

Description Small round-headed warbler with short wings and stout, short, dark bill with yellow at base of lower mandible. Olive-brown upperparts, white throat and light-brown belly. Prominent supercilium is wider and buff-coloured in front of eye becoming whiter above and behind eye.
Voice Call is a short, harsh and slightly muted *tuk-tuk*.
Similar species Dusky Warbler has greyer upperparts and a thinner bill; supercilium is wider and buff-coloured behind eye.

Kazuyasu Kisaichi

First-winter. October, Tobishima

Range Breeds from Siberia east to Sakhalin and Ussuriland, south to northern China and Korea. Winters in south-east Asia.
Status in Japan Rare passage migrant to thickets and reed beds. Skulks in undergrowth and difficult to observe. Mostly recorded on offshore islands in the Sea of Japan.

Yoshihito Goto

First-winter. September, Hegura-jima

Ran Schols

Adult. May, Beidaihe, China

Pallas's Leaf Warbler

Phylloscopus proregulus 10–10.5cm

Ran Schols

Adult. May, Bedaihe, China

Description Tiny, compact warbler, greenish-brown above with yellow crown-stripe and pale-yellow supercilium extending to forehead, two yellowish wing-bars, white edges to the tertials, a yellow rump and whitish underparts.

Voice Call is a short *tchii.*

Similar species Separated from other warblers by tiny size, and the yellow crown-stripe and rump.

Norio Yamagata

Adult. May, Hegura-jima

Range Breeds from Siberia east to Ussuriland, Sakhalin and northern China, also in the Himalayas. Winters from India east to southern China and northern part of south-east Asia.

Status in Japan Very rare passage and winter visitor to wooded areas. Mostly recorded on islands in the Sea of Japan.

Ran Shoals

Adult. May, Bedaihe, China

Arctic Warbler

Phylloscopus borealis 12.5–13cm

Description Greenish-brown above with a yellowish supercilium, dark eye-stripe and mottled ear-coverts. Also a single pale wing-bar, whitish underparts with variable amount of yellow wash, dark bill with yellow lower mandible and brown legs.
Voice Song is a monotonous, fast whirring thrill *tsiri siri siri siri siri siri,* call is a metallic *tzi.*
Similar species Sakhalin Leaf Warbler has a dark greyish-brown crown, pinkish legs and different call. Eastern Crowned Warbler has pale crown-stripe. See also Greenish Warbler.

Tadao Shimba

First-winter. October, Hegura-jima

Range Breeds across northern Eurasia from Scandinavia east to far north-eastern Russia and western Alaska, south to Sakhalin and Japan. Winters in south-east Asia.
Status in Japan Common summer visitor to montane coniferous forests on Hokkaido, Honshu and Kyushu. Some occur in wooded urban areas and parks during migration. Arrives by late May and departs by October, later than other migrant warblers.

Tadao Shimba

First-winter. October, Hegura-jima

Akira Yamamoto

Yellow-browed Warbler

Phylloscopus inornatus 11–12cm

Teruaki Ishii

Adult. May, Hegura-jima

Kazuyasu Kisaichi

First-winter. November, Tobishima

Description Small warbler, greenish-brown above with a pale supercilium and two wing-bars, broad white fringes to the tertials and secondaries and whitish underparts.

Voice Call is a slurred *cheweeh,* often first detected by call.

Similar species Pallas's Leaf Warbler has a yellow crown-stripe and yellow rump.

Range Breeds from Siberia east to the Okhotsk coast, south to Ussuriland and northern China. Winters in Taiwan, southern China and northern part of south-east Asia.

Status in Japan Rare passage migrant to wooded areas and grassy fields with thickets. Moves quickly and nervously from tree canopy to understorey. Regularly seen on offshore islands in the Sea of Japan. Some winter on the Ryukyu Islands.

Shigeo Ozawa

Sakhalin Leaf Warbler

Phylloscopus borealoides 11–12cm

Description Dark greyish-brown crown, white supercilium, dark eye-stripe and mottled ear-coverts. Greenish-brown above with a single whitish wing-bar when in fresh plumage, whitish underparts and pinkish legs.
Voice Song is a thin, metallic and slow trisyllabic *hee-tooh-keeh*.
Call is a metallic *tink*.
Similar species Pale-legged Leaf Warbler is almost identical and best separated by song and range. Arctic Warbler is uniform greenish-brown above and has a different song and call. Eastern Crowned Warbler has a pale yellowish crown-stripe.

Range Breeds from Sakhalin south to Japan and winters in south-eastern Asia.
Status in Japan Common summer visitor to montane mixed forests from central Honshu northwards. some occur in wooded urban areas and parks during migration.

Teruaki Ishii

Adult. May, Hegura-jima

Tadao Shimba

Adult. May, Hegura-jima

Kiyoto Ogata

Juvenile. September, Aichi

Eastern Crowned Warbler

Phylloscopus coronatus 12–13cm

Akira Yamamoto

Adult. April, Aichi

Description Greenish-brown above with a pale crown-stripe and white supercilium. Has a dark eye-stripe, mottled ear-coverts, dark bill with yellow base to lower mandible, whitish underparts with yellow wash on lower belly and a single whitish wing-bar.
Voice Song is loud and distinctive *tswui-tsui tswuee*. Call is a soft *phit-phit*.
Similar species Arctic Warbler has longer wings, lacks a crown-stripe and has a different song and call. Ijima's Leaf Warbler lacks a crown-stripe and has a different song.

Mitsuo Imai

Adult. May, Hegura-jima

Range Breeds in eastern Asia from south-eastern Transbaikalia east to Amur basin and Ussuriland, south to north-eastern China and Korea. Winters in south-east Asia.
Status in Japan Common summer visitor to deciduous forests from Kyushu northwards.

Shuichi Tsuno

Juvenile. September, Wakayama

Ijima's Leaf Warbler

Phylloscopus ijimae 12cm

Description Greenish-brown above with a yellowish supercilium, dark eye-stripe and white eye-ring, dark bill with yellow base to lower mandible and whitish underparts.
Voice Song is loud and high pitched *chow-chow-chow-chow.*
Similar species Eastern Crowned Warbler has a pale crown-stripe, broader supercilium and a different song.

Tadao Shimba

Adult. June, Miyake-jima

Range Breeds exclusively on the Izu and Tokara Islands in Japan and winters in the Philippines.
Status in Japan Locally common summer visitor to evergreen forests, very rare passage migrant elsewhere. There are a few winter records from the Ryukyu Islands.

Tadao Shimba

Adult. June, Miyake-jima

Tadao Shimba

Adult. June, Miyake-jima

Greenish Warbler

Phylloscopus trochiloides 11cm

Ran Schols

Adult. May, Bedaihe, China

Description Greenish-brown above with a prominent yellowish supercilium and whitish underparts. Has two yellowish wing-bars; bill has dark upper mandible, pale lower mandible. Legs grey-brown.
Voice Call is a shrill *chee-wee* or *tiss-yip*.
Similar species Arctic Warbler is slightly larger with a narrower and longer supercilium, longer wings, broader eye-stripe and dark tip to lower mandible.

Range Breeds across northern Eurasia east to Okhotsk coast, Ussuriland and north-eastern China, south to central Asia and central China. Winters in southern Asia. Rare passage migrant to Korea.
Status in Japan Very rare passage migrant to offshore islands in Sea of Japan. All records refer to the race *plumbeitarsus*, considered by some authorities to be a separate species: Two-barred Greenish Warbler.

Asian Brown Flycatcher

Muscicapa dauurica 13cm

Mitsuo Imai

Adult. May, Hegura-jima

Description Greyish-brown flycatcher. Sexes alike. White eye-ring and lores, grey malar stripe, white underparts with pale-grey wash on breast and flanks. Juvenile has white spots on breast and belly and pale edges to wing-coverts.
Voice Song is squeaky, high-pitched, melodious chattering. Call is a thin, metallic *seeh*.
Similar species Dark-sided Flycatcher has a dark-grey breast. Grey-streaked Flycatcher has distinctive heavy streaks on breast and flanks.

Tadao Shimba

Adult. May, Hegura-jima

Range Breeds in southern Siberia from the Yenisei east to the Amur Basin, Ussuriland and Sakhalin, south to northern Korea and Japan, also in the Himalayas and the Philippines. Northern populations are migratory and winter south to southern Asia.
Status in Japan Common summer visitor to broad-leaved deciduous forests from Kyushu northwards.

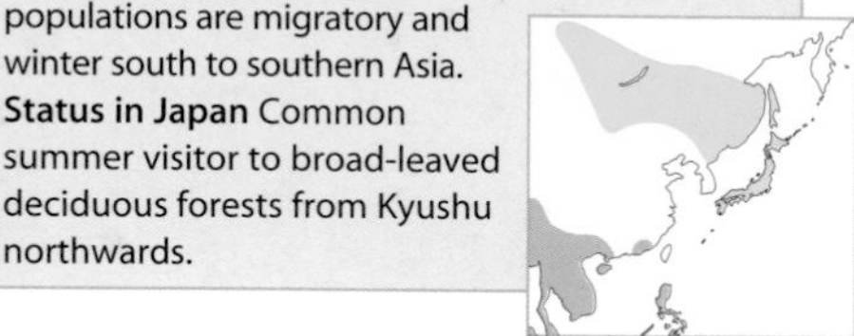

Dark-sided Flycatcher

Muscicapa sibirica 13.5cm

Other name Siberian Flycatcher
Description Dark, greyish-brown back, whitish underparts with dark-grey breast and white eye-ring. Sexes alike. Juvenile has buff-coloured edge to wing-coverts.
Voice Song is squeaky, high-pitched, melodious chattering and similar to Asian Brown Flycatcher. Call is a sharp, thin *tseeh*.
Similar species Asian Brown Flycatcher has pale lores and underparts. Grey-streaked Flycatcher has distinctive dark streaks on breast and belly.

Mitsuo Imai
Adult. May, Hegura-jima

Range Breeds in southern Siberia from the Altai Mountains east to the Okhotsk coast and Kamchatka, south to north-east China and northern Japan, also in the Himalayas. Winters in south-east Asia.
Status in Japan Uncommon summer visitor to coniferous forests from central Honshu northwards; uncommon passage migrant elsewhere. Breeds in subalpine parts of Honshu.

Grey-streaked Flycatcher

Muscicapa griseisticta 14.5cm

Description Dark greyish-brown above, long wings with white edges to greater-coverts and white underparts with heavy dark streaks on breast and flanks.
Voice Call is sharp, thin *tseeh*.
Similar species Asian Brown Flycatcher has pale unstreaked underparts and shorter wings. Dark-sided Flycatcher has a darker grey breast.

Tadao Shimba
First-winter. September, Mie

Range Breeds in far-eastern Asia from the Kamchatka Peninsula south to Sakhalin and north-eastern China. Winters in south-east Asia. Common passage migrant to Korea.
Status in Japan Passage migrant in wooded areas. Rare in spring and common in autumn.

Mitsuo Imai
First-winter. September, Hegura-jima

Yellow-rumped Flycatcher

Ficedula zanthopygia 13cm

Akira Yamamoto

Adult ♂ breeding. May, Tobishima

Description Male has a black head with distinctive white supercilium, black back, tail and wings with a white patch on inner greater coverts, yellow rump and yellow underparts. Female is olive-brown above and pale brown below with a similar white wing patch, yellow rump and pale yellow throat. **Voice** Similar to Narcissus Flycatcher. **Similar species** Male Narcissus Flycatcher has yellow supercilium and orange throat. Female Narcissus Flycatcher has olive-brown rump and lacks white wing patch.

Shigeo Ozawa

First-summer ♂. May, Hegura-jima

Range Breeds from eastern Mongolia east to Amur Basin and Ussuriland, south to Korea and eastern China. Winters in South-east Asia.
Status in Japan Rare passage migrant to wooded areas. Frequently seen on offshore islands in Sea of Japan in May.

Shigeo Ozawa

♀. May, Hegura-jima

Narcissus Flycatcher

Ficedula narcissina 13.5cm

Description Male has black head with yellow supercilium, black back, tail and wings with a white patch on wing-coverts and a yellow rump. Orange throat, yellow breast and white belly and undertail-coverts. Female has olive-brown head, back and tail and pale throat; white underparts with pale-brown wash on breast. Two races in Japan; *owstoni* on Ryukyu Islands is slightly smaller and has a greenish-black back.
Voice Melodious song of repeated warbles and whistles. Call is a metallic, monotonous and repeated *pwit-pwit*.
Similar species Female Blue-and white Flycatcher is larger with brown breast. Female Yellow-rumped Flycatcher has a white patch on wing-coverts and yellow rump.

Tokio Sugiyama
Adult ♂ breeding, May, Hegura-jima

Tadao Shimba
First-summer ♀. May, Hegura-jima

Range Breeds in Sakhalin and Japan; also in northern China. Winters in southern China and south-east Asia.
Status in Japan Nominate *narcissina* is a common summer visitor to deciduous forests from Kyushu northwards; *owstoni* is an uncommon resident in evergreen forests of Ryukyu Islands.

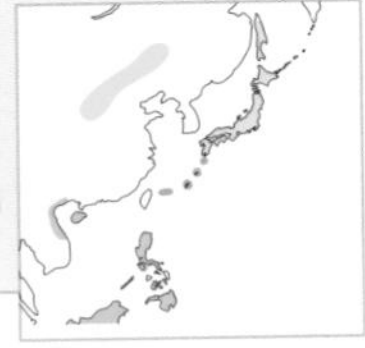

Tadao Shimba
First-summer ♂. May, Hegura-jima

Mugimaki Flycatcher

Ficedula mugimaki 13cm

Mitsuo Imai

Adult ♂ breeding, May, Hegura-jima

Mitsuo Imai

First-summer ♂. May, Hegura-jima

Description Male has black head with white patch behind eye, black back and wings with white distinctive shoulder patch and white edges to tertials. Also a bright orange throat, breast and belly, white lower belly and undertail-coverts and black tail with white sides at the base. Female has a greyish-brown head, back and tail, pale orange throat and breast, and white belly. First–year male is similar to female but has a pale spot behind eye and white at the base of tail.
Voice Call is a dry *tyut-tyut*.
Similar species Female Narcissus Flycatcher lacks orange wash on breast.

Range Breeds in southern Siberia east to the Amur Basin, Ussuriland and Sakhalin, south to north-eastern China. Winters in southern China and south-east Asia.
Status in Japan Rare passage migrant to wooded areas. Regularly seen on offshore islands in the Sea of Japan.

Akira Yamamoto

Adult ♀. May, Hegura-jima

Taiga Flycatcher

Ficedula albicilla 11–12cm

Description Small flycatcher with a prominent white base to outer feathers of the black tail. Male has grey face, orange throat, dark greyish-brown back and greyish underparts. Female and immature male have greyish-brown face and back, white eye-ring and white underparts. Tail is frequently cocked.
Voice Call is buzzing *zizit-zizit*.
Similar species Asian Brown Flycatcher lacks white patches on the tail.

Akira Yamamoto
Adult ♂. December, Aichi

Range Breeds from the Urals east to the Okhotsk coast and Kamchatka, south to Transbaikalia and lower Amur Basin. Winters south to southern China and the Malay Peninsula.
Status in Japan Rare passage migrant and winter visitor to wooded areas, city parks and gardens.

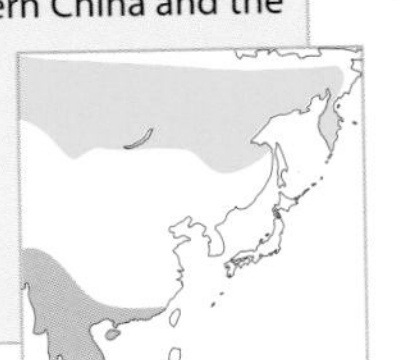

Manabu Totsuka
First-winter, March, Aichi

Pied Flycatcher

Ficedula hypoleuca 13cm

Description Male is sooty-black above with a white spot on forehead, white wing patch and black tail with white outer-tail feathers. Female has a greyish-brown back, white wing patch and a black tail with white outer-tail feathers.

Range Breeds in northern Eurasia from northern Europe east to the Yenisey. Winters in tropical Africa.
Status in Japan Vagrant. Recorded on Hegura-jima (1991) and Miyagi (2006).

Mitsuo Imai
First-winter ♀. October, Miyagi

Blue-and-white Flycatcher

Cyanoptila cyanomelana 16–16.5cm

Akira Yamamoto

Adult ♂. May, Hegura-jima

Description Male has pale glossy-blue forehead and crown, blue back and tail; black face and breast and white underparts. Female has a brown head, back and breast, reddish-brown tail and white belly.
Voice Song is a loud and melodious warble mixed with distinctive call-like notes, *zit-zit*. Delivered from exposed perch, often high up in trees. Call is *huit-huit*.
Similar species Female Narcissus Flycatcher is smaller with paler throat and brown tail.

Yoshio Goto

Adult ♀. May, Hegura-jima

Range Breeds in far-eastern Asia from Amur Basin and Ussuriland, south to eastern China. Winters south to south-east Asia. Common summer visitor to Korea.
Status in Japan Common summer visitor to deciduous and mixed forests near streams and rivers from Kyushu northwards.

Tadao Shimba

Juvenile ♂. October, Hegura-jima

Japanese Robin

Luscinia akahige 14cm

Description Male has a dark-orange face and breast, reddish-brown back and tail and greyish-white underparts with a dark breast-band. Izu Islands race *tanensis* lacks dark breast-band. Female is duller and lacks breast-band.
Voice Song is a single high note followed by a rattle; *chjuui rarararararara*. Call is a thin metallic *tsip*.
Similar species Ryukyu Robin has rusty-orange back and is mainly restricted to the Amami and Okinawa Islands.

Range Breeds in southern Sakhalin and Japan; winters in southern China.
Status in Japan Common summer visitor to montane mixed and deciduous forests from Kyushu northwards. Skulks on the forest floor and is often difficult to observe. Race *tanensis* is uncommon resident on the Izu Islands, Yakushima and Tanegashima.

Akira Yamamoto

Adult ♂ *akahige*. May, Nara

Tadao Shimba

Adult ♀ *akahige*. May, Hegura-jima

Tadao Shimba

Adult ♂ *akahige*. May, Hegura-jima

Ryukyu Robin

Luscinia komadori 14cm

Kazuyasu Kisaichi

Adult ♂ *namiyei*. September, Okinawa-Hontou

Kazuyasu Kisaichi

Adult ♀ *namiyei*. September, Okinawa-Hontou

Description Male has distinctive black face, throat and breast, rusty-orange upperparts and white underparts and dark bill and legs. Female has rusty-orange face and back, whitish underparts with dark spots on breast and flanks. Two races: male of *komadori* on Osumi and Amami Islands has distinctive black patch on flanks; male *namiyei* on Okinawa Islands has dark-grey underparts and lacks dark flank patch.
Voice Song is a melodious, penetrating series of notes, often beginning as rapid trill followed by *tu-tee, tu-tee, tu-tee, tu-tee.*

Range Endemic to southern Japan.
Status in Japan Locally common resident on offshore islands from south of Kyushu to the Okinawa Islands. Northern population moves south in winter. Skulks on the floor of dense sub-tropical forests.

Tokio Sugiyama

Adult ♂ *komadori*. February, Amami-Oshima

Rufous-tailed Robin

Luscinia sibilans 13cm

Description Uniform brown above with a reddish-brown tail, dark bill, pale lores and white eye-ring. Whitish below with distinct, scaly, greyish-brown marks on breast and belly.
Voice Song consists of long trilling notes, repeated for several seconds on a downward scale.
Similar species Female Siberian Rubythroat is slightly larger and has a white supercilium. Female Siberian Blue Robin is paler, has bluish tail and lacks scaly marks on breast.

Range Breeds in the boreal zone of Siberia east to Ussuriland, Sakhalin and north-eastern China. Winters in southern China and northern parts of south-east Asia.
Status in Japan Rare spring passage migrant to offshore islands in the Sea of Japan.
Secretive and skulks in undergrowth. Often detected by song delivered from dense cover.

Tadao Shimba

First-summer. May, Hegura-jima

Tadao Shimba

Adult. May, Hegura-jima

Tadao Shimba

Adult. May, Hegura-jima

Red-flanked Bluetail

Luscinia cyanura 14cm

Tadao Shimba

Adult ♂. February, Aichi

Description Male has bright-blue head and upperparts, white supercilium and dark wings. White throat and underparts with bright yellowish-orange flanks, and a bluish tail. Female has olive-brown upperparts, bluish tail, brown breast and a white belly with yellowish-orange flanks.
Voice Melodious song, starting with trilling *prui-churrii chichichir*.
Similar species Female Daurian Redstart has a rufous tail and distinctive white wing-patch.

Tadao Shimba

Adult ♀. May, Hegura-jima

Range Breeds across Eurasia from North-eastern Europe through southern Siberia to Kamchatka, Sakhalin and northern China. Winters from the Indian subcontinent east to southern China and northern south-east Asia.
Status in Japan Common resident. Breeds in mixed subalpine forests from Shikoku northwards and winters from Honshu southwards. Often seen in city parks and foothills in winter.

Tadao Shimba

Second-winter ♂. February, Shizuoka

Bluethroat

Luscinia svecica 15cm

Description Male has olive-brown upperparts, white supercilium, chestnut ear-coverts and a striking blue throat with orange central spot. Whitish underparts with black, white and orange breast-bands and a dark-brown tail with orange outer base. Female has a white moustachial stripe and white throat bordered black.
Similar species Female Siberian Rubythroat is similar to female, but lacks dark streaks on breast and orange tail patches.

Kazuyasu Kisaichi

Adult ♂ breeding. March, Osaka

Range Breeds across northern Eurasia from Europe east through central Asia to north-eastern Russia and western Alaska. Winters from southern Europe and northern Africa east to south-east Asia.
Status in Japan Rare passage migrant and winter visitor to cultivated fields with thickets, riverbanks and reedbeds.

Akira yamamoto

Adult ♀. April, Aichi

Mitsuo imai

First-winter ♂. February, Kanagawa

Siberian Blue Robin

Luscinia cyane 14cm

Akira Yamamoto

Adult ♂. May, Hegura-jima

Akira Yamamoto

Adult ♀. May, Hegura-jimaa

Description Male has dark-blue upperparts contrasting with white underparts; black lores, ear-coverts and sides of neck, flanks bluish. Female has olive-brown upperparts, bluish tail, and pale lores; buff-coloured below with faint, scaly, brownish marks on breast.
Voice Song contains a quickly repeated metallic prelude *tsit-tsit-tsit,* followed by loud melodious warbles. It is delivered from dense undergrowth; birds are difficult to observe.
Similar species Rufous-tailed Robin is similar to female, but has rufous tail and more distinct scaly greyish-brown marks on the breast and belly.

Range Breeds from southern Siberia east to Sakhalin, south to north-eastern China, Korea and Japan. Winters from India east to south-east Asia.
Status in Japan Common summer visitor to montane deciduous forests with dense understorey from central Honshu northwards. Skulks on the forest floor and prefers moist areas.

Tokio Sugiyama

First-summer ♂. May, Hegura-jima

Siberian Rubythroat

Luscinia calliope 15–16cm

Description Male has olive-brown upperparts, white supercilium, dark lores, white moustachial stripe, a distinctive bright-red throat, greyish-brown breast and flanks and a pale belly. Female is paler with white throat.
Voice Song is a variety of melodious whistles; call is a loud whistle *feeyou-feeyou* or harsh *chak*.
Similar species Rufous-tailed Robin is similar to female but smaller, and lacks white supercilium.

Tokio Sugiyama

Adult ♂ breeding. May, Hegura-jima

Range Breeds in Siberia east to Kamchatka, south to northern China, Sakhalin and northern Japan. Winters in southern Asia.
Status in Japan Common summer visitor to Hokkaido, from coastal grasslands to subalpine zone. Some also breed in northern Honshu. Uncommon passage migrant elsewhere; small numbers regularly winter on the Ryukyu Islands.

Yoshio Goto

First-summer ♀. May, Hegura-jima

Black Redstart

Phoenicurus ochruros 14cm

Description Male has black head and breast, greyish crown and back, rusty-red underparts and rusty-red uppertail-coverts and outer tail feathers. Female is greyish-brown overall with pale brown undertail-coverts and rufous outer tail feathers.
Similar species Female Common Redstart is paler and browner. Female Daurian Redstart is browner, has a white wing patch and chestnut undertail-coverts.

Range Breeds from Europe east to Mongolia and western China. Winters from southern Europe and northern Africa, east to the Indian subcontinent.
Status in Japan Very rare passage migrant to cultivated and grassy fields. Mostly recorded in early spring.

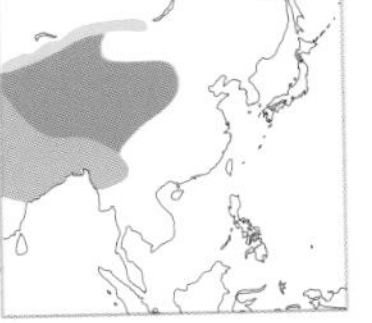

Kenji Takehara

First-summer ♂. March, Kume-jima

Daurian Redstart

Phoenicurus auroreus 14cm

Tadao Shimba

Adult ♂ breeding. March, Aichi

Description Male has a black face, silvery-grey crown and hindneck, dark-brown back, a conspicuous white wing-patch, and chestnut underparts, rump and uppertail-coverts. Female has a greyish-brown face, breast and back, and chestnut lower belly and undertail-coverts.

Voice Song is a loud, melodious rambling warble. Call is a short and metallic *pyut-pyut* or harsh *tac-tac*.

Similar species Female Black Redstart is greyer and lacks white on wings. Female Common Redstart has pale chestnut-tinged brown breast and belly, and also lacks white on wings.

Manabu Totsuka

Adult ♀. March, Aichi

Range Breeds from southern Siberia east to Sakhalin and Ussuriland, south to Korea and northern China. Winters from north-eastern India east to southern China and south-east Asia.

Status in Japan Common winter visitor to foothills, cultivated fields and wooded urban areas. Arrives during October, departs by April; a single breeding record from Hokkaido. Often feeds on *Pyracantha* in gardens.

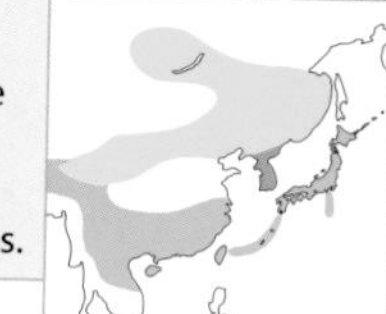

Tadao Shimba

First-winter ♂. March, Aichi

Common Redstart

Phoenicurus phoenicurus 14cm

Description Male has a black face, silvery-white forehead and supercilium, grey back, chestnut rump, chestnut tail with dark brown central feathers, dark-brown wings and chestnut underparts. Female has greyish-brown upperparts, a pale throat, and chestnut-tinged brown breast and belly.
Similar species Female Black Redstart is greyer overall, has darker throat and belly. Female Daurian Redstart is browner and has white on wings.

Range Breeds from Europe east to about Lake Baikal. Winters mainly in tropical Africa and the Arabian Peninsula.
Status in Japan Vagrant. First recorded in Nov. 1998 on Hegura-jima. Two other sightings on Hegura-jima subsequently.

Norio Yamagata

Adult ♂ non-breeding. November, Hegura-jima

Pied Bushchat

Saxicola caprata 14cm

Description Male is black with white patches on wing-coverts, white rump and white lower belly and undertail-coverts. Female is greyish-brown with darker brown wings and tail and rusty uppertail-coverts.
Similar species Female Grey Bushchat has an olive-brown back, white supercilium and a white throat.

Range Breeds from eastern Iran east through Indian subcontinent to south-east Asia.
Status in Japan First recorded in 1989 on Yonaguni-jima; another sighting there in 2004. Occurs in cultivated and grassy fields.

Himaru Iozawa

First-winter ♂. January, Yonaguni-jima

Siberian Stonechat

Saxicola maurus 13cm

Manabu Totsuka

First-summer ♂. June, Hokkaido

Description Male has black hood, back and tail, white rump and underparts and orange-brown breast. First-summer male has whiter breast. Female has brown head and back with dark streaks, pale supercilium and orange-brown breast, dark tail and buff-coloured underparts and rump.
Voice Song is a variety of twittering trills. Call is a harsh *tzit-tzit*.
Similar species Female Whinchat *S. rubetra* (recorded once on Okinawa-Hontou) has a distinct supercilium and dark streaks on rump.

Tadao Shimba

Adult ♀. June, Nagano

Range Breeds from central Siberia to north-eastern Siberia and Sakhalin. Northern populations are migratory.
Status in Japan Common summer visitor to central Honshu northwards; common passage migrant on cultivated and grassy fields from Honshu southwards. Breeds in highland meadows and grassy fields with thickets. Arrives during April, departs by October.

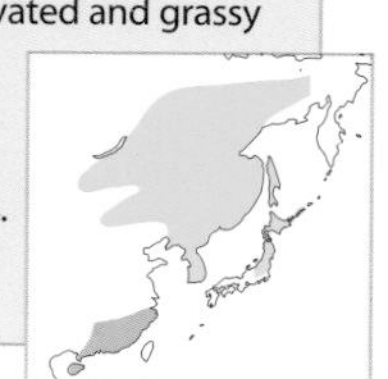

Tokio Sugiyama

First-winter ♀. October, Aichi

Grey Bushchat

Saxicola ferrea 15cm

Description Male has grey crown and mantle with dark streaks, whitish underparts. Also dark wings with a small white patch, dark tail, white super cilium, black lores and black ear-coverts. Female has brown upperparts, pale supercilum, white throat, orange-tinged brown breast and belly and white undertail-coverts.

Teruaki Ishi

First-winter ♂. September, Hegura-jima

Range Breeds from the Himalayas east to southern China and northern south-east Asia. Partially migratory.
Status in Japan Vagrant. Recorded on Hegura-jima, Yakushima, Iriomote-jima and Yonaguni-jima. Occurs in cultivated and grassy fields.

Pied Wheatear

Oenanthe pleschanka 14.5cm

Description Male breeding has a black face, throat and upperparts, silvery-white crown and hindneck; light buff-coloured underparts, white rump, white tail with black terminal band, and mainly black central feathers. Female is greyish-brown above and has whitish underparts with buff-coloured wash on breast.
Similar species Non-breeding and juvenile Northern Wheatear is browner, and has buff-coloured underparts. Desert Wheatear is also browner and has a different tail pattern.

Shigeo Ozawa

First-summer ♂. May, Tsushima

Range Breeds from central Asia east through southern Russia to Mongolia and north-western China. Winters mainly in eastern Africa.
Status in Japan Rare visitor to cultivated and stony fields. Mostly recorded around the coasts of the Sea of Japan during spring and autumn migration.

Mitsuo Imai

First-winter. September, Hegura-jima

Northern Wheatear

Oenanthe oenanthe 14.5cm

Mitsuo Imai

Adult ♂ breeding. May, Tsushima

Mitsuo Imai

First-winter. September, Hegura-jima

Description Male breeding has bluish-grey crown and mantle, white supercilium and distinctive black eye-stripe and ear-coverts. White underparts with buff-coloured wash on breast; black wings, white rump and uppertail-coverts, white tail with black terminal band and mainly black central feathers. Female is browner, with brown eye-stripe and ear-coverts.

Similar species Isabelline Wheatear is similar to female and non-breeding, but slightly larger and has thicker bill, longer legs, and broader terminal band on tail.

Range Breeds across northern Eurasia, Greenland and western Alaska. Winters mainly in Africa.

Status in Japan Rare passage migrant on cultivated and grassy fields.

Mitsuo Imai

First-winter. September, Hegura-jima

Desert Wheatear

Oenanthe deserti 14.5cm

Description Male has black face and throat, sandy-brown upperparts, black wings with pale edges, white rump and almost entirely black tail. Female lacks black on head and is grey-brown above. Tail largely black.
Similar species Separated from other wheatears by mainly black tail.

Tadao Shimba

Adult ♂ non-breeding. January, Tokyo

Range Breeds from central Asia east to Mongolia and eastern China. Winters mainly in east Africa and the Arabian Peninsula.
Status in Japan
Rare winter visitor on stony fields and riverbanks. Feeds mainly on the ground.

Shigeo Ozawa

First-winter ♀. February, Tokyo

Tadao Shimba

Adult ♂ non-breeding. January, Tokyo

Isabelline Wheatear

Oenanthe isabellina 16cm

Tokio Sugiyama

Adult ♂ breeding. April, Tsushima

Description Sandy-brown upperparts tinged rusty, indistinct pale supercilium and dark brown wings with buff-coloured edges and contrasting black alula. Buff-coloured underparts, white rump and uppertail-coverts and white tail with broad black terminal band. Sexes similar but male has darker lores.
Similar species Northern Wheatear is smaller, with shorter legs, thinner bill and narrower black terminal band on tail.

Manabu Totsuka

Adult ♀. April, Hegura-jima

Range Breeds from Turkey east through central Asia to southern Siberia, Mongolia and northern China. Winters mainly in tropical Africa and the Arabian Peninsula.
Status in Japan Rare visitor to cultivated and grassy fields. Mostly recorded in April-May.

Tokio Sugiyama

Adult ♂ breeding. April, Tsushima

Japanese Paradise Flycatcher

Terpsiphone atrocaudata ♂44.5cm ♀17.5cm

Description Long-tailed flycatcher. Male has very long, dark tail, purplish-black head, breast and upperparts and a bright-blue eye-ring and bill. Short crest and white belly. Female has sooty-black head and breast, dark reddish-brown upperparts and tail, a white belly, short crest and a blue eye-ring and bill.
Voice Song often begins with slightly buzzing notes, followed by loud, piercing whistles. Very distinctive and often first detected by song. Call is a *chee-tewi*.

Range Breeds in Japan and Korea; also on offshore islands in Taiwan and in northern Philippines. Winters in south-east Asia.
Status in Japan
Uncommon summer visitor to Honshu southwards; common on the Ryukyu Islands. Breeds in dense forests in foothills.

Norio Yamagata

Adult ♂. July, Shizuoka

Kazuyasu Kisaichi

Adult ♀ *illex*. July, Miyako-jima

Norio Yamagata

Adult ♀. July, Shizuoka

Asian Paradise Flycatcher

Terpsiphone paradisi ♂48cm ♀21cm

Michelle & Peter Wong

Adult ♀, rufous phase. October, Hong Kong

Description Long-tailed passerine with blue bill and eye-ring and a short crest. The very long-tailed male has two distinct colour phases, white and rufous. White male is largely white with a glossy black head and throat. Rufous male has black head and throat, rufous back and tail, grey breast and white belly. Female is similar to rufous male but crest and tail shorter.

Range Breeds from Turkestan east through Indian subcontinent to south-east Asia; north to north-east China and Ussuriland. Northern populations are migratory and winter in southern Asia. Inhabits dense forests.

Bearded Reedling

Panurus biarmicus 16cm

Tomi Muukkonen

Adult ♂. February, Espoo, Finland

Description Long tailed, tit-like bird of reedbeds. Male has grey head, small yellow bill, distinctive long black moustaches and rufous-brown back, tail and flanks. Has dark wings with largely white primaries, pale belly and black undertail-coverts. Female has plain buff-coloured head and lacks black moustaches.

Range Breeds from western Europe east through Siberia and Mongolia to Ussuriland and north-eastern China.

Status in Japan Vagrant. Several records in Honshu.

Vinous-throated Parrotbill

Paradoxornis webbianus 12cm

Description Small, mainly rufous-brown bird with a small conical bill, long tail, reddish-brown wings and buff-coloured underparts.

Tokio Sugiyama

Adult. December, Pusang, South Korea

Range Eastern Asia from Ussuriland south through eastern China and Korea to southern China. Common resident in Korea.
Status in Japan Vagrant. Recorded once in Niigata in 1984. Forms feeding flocks in winter. Occurs in grassy fields and riverbanks with thickets.

Reed Parrotbill

Paradoxornis heudei 18cm

Description Long-tailed and sparrow-sized with yellowish, downcurved, parrotlike bill. Greyish-white head, nape and breast with diagnostic broad dark-brown eyeline extending to the nape; rusty-brown back and underparts.
Voice Call includes short trills and nasal whistles.

Yuri Shibnev

Adult at nest. Lake Khanka, Ussuriland

Range Breeds in reedbeds from Lake Khanka in Ussuriland south to Zhejiang in eastern China. Secretive. Threatened, and declining due to habitat destruction.
Status in Japan No records.

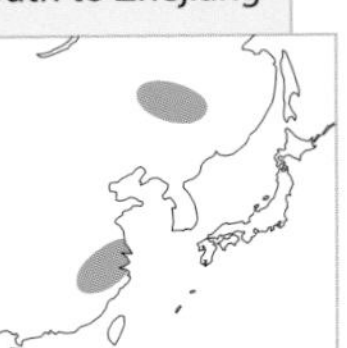

Long-tailed Tit

Aegithalos caudatus 13cm

Tadao Shimba

Adult *trivirgatus*. February, Nagano

Description Small, whitish, long-tailed tit. Broad, black stripe above eye. Pinkish scapulars, flanks and belly; dark back, wings and tail. Tail has white outer feathers. Four races in Japan; *japonicus* in Hokkaido has all-white head, others almost identical.
Voice Thin, metallic song. Contact call *jurrrh-jurrrh* is frequently given by feeding flocks.

Tadao Shimba

Adult *trivirgatus*. February, Aichi

Range Resident in temperate Eurasia from Europe east to far-eastern Russia, Sakhalin, Korea, eastern China and Japan.
Status in Japan Common resident in wooded areas from Kyushu northwards. In winter, forms flocks that are often associated with other tits, Eurasian Nuthatch, Japanese White-eye and Japanese Pygmy Woodpecker.

Yoshihito Goto

Adult *japonicus*. January, Hokkaido

Coal Tit

Periparus ater 10.5–11cm

Description Small tit with short tail, two white wing-bars and a white nape patch. Black crown and throat, short crest, white cheeks, bluish-grey above and buffish below.
Voice Song is a high-pitched *chupih-chupih-chupih*; contact call is a thin *si-si* or *trrrh.*

Tadao Shimba

Adult *insularis.* January, Hokkaido

Range Widespread across northern Eurasia from Europe east to far-eastern Russia, north-eastern China and Korea.
Status in Japan Common resident in montane coniferous forests from Kyushu northwards. Moves to lower altitudes in winter.

Tim Edelsten

Adult *ater*. February, Ilsan, South Korea

Tadao Shimba

Adult *insularis.* February, Yamanashi

Willow Tit

Poecile montanus 12.5cm

Tadao Shimba

Adult. January, Yamanashi

Description Dull black cap, small black bib, greyish-black back, dark-brown wings with white edges to primaries. White cheeks and underparts.
Voice Song is a repeated *tju-pi tju-pi* or *pju-pju pju-pju pju-pju*. Call is *zi-zi tjaa tjaa tjaa*.
Similar species Marsh Tit has stouter bill, larger bib and a different call. Songer Tit has cinnamon-washed underparts.

Tadao Shimba

Adult. February, Yamanashi

Range Widespread across northern Eurasia from Europe east to far-eastern Russia, north-eastern China and Korea.
Status in Japan Common resident in mixed montane forests from Kyushu north-wards. Inhabits lower altitudes on Hokkaido.

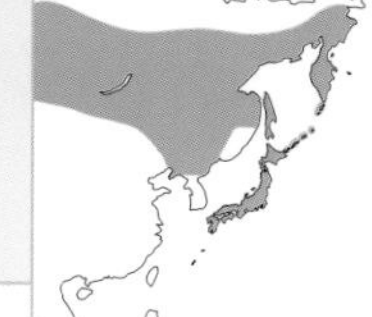

Songar Tit

Poecile songarus 13cm

Lars Hansson

Adult. July, Qinghai, China

Description A counterpart of Willow Tit in central Asia and China. Cap and large bib dull blackish-brown. Cheek patches off-white. Pale cinnamon above with paler wing panel; underparts cinnamon-buff.
Voice Similar to that of Willow Tit but more varied, and calls are more rasping.
Similar species Willow Tit has white underparts. Marsh Tit lacks pale wing-panel.

Range Locally common in montane coniferous forests in the Tien Shan and in south-west China; also in north-central China. Uncommon resident in mixed forests in north-east China north to about Beijing and Hebei.
Status in Japan No records.

Marsh Tit

Poecile palustris 12.5cm

Description Glossy-black cap, black bib, pale greyish-brown back with plain dark-brown wings. White cheeks and underparts.
Voice Song is a repeated *chju chju chju chju chju*; call is an explosive *piitjee*.
Similar species Willow Tit has thinner bill and smaller bib, white edges to primaries, and a different call.

Tim Edelsten

Adult *hellmayri*. February, Seoul, South Korea

Range Resident in western and eastern Eurasia. Eastern range from Altai east to Sakhalin, south to north-eastern China and Korea.
Status in Japan Common resident in broad-leaved deciduous forests on Hokkaido.

Tadao Shimba

Adult *hensoni*. January, Hokkaido

Igor Karyakin

Adult *brevirostris*. August, Ussuriland, Russia

Varied Tit

Poecile varius 14–15cm

Tadao Shimba

Adult *varius*. January, Shizuoka

Description Unmistakable. Black head and breast, pale yellowish-brown forehead and face. Orange-brown back and underparts, dark bluish-grey upperparts and tail. Several races in Japan; southern birds are darker. Race *owstoni* on the Izu Islands is darkest and has dark reddish-brown face and underparts and a larger bill. **Voice** Song is slow pitched, rising, nasal *tzee-tzee-wheee*. Contact call is a nasal *bee-bee*.

Tadao Shimba

Adult *varius*. February, Yamanashi

Range Southern Korea, Japan and Taiwan.
Status in Japan Common resident in deciduous and evergreen forests throughout Japan. Uncommon on Hokkaido.

Tadao Shimba

Adult *owstoni*. June, Miyake-jima

Siberian Tit

Poecile cincta 14cm

Description Brown cap and back, white cheeks and black bib. Greyish-brown wings and tail and white underparts with buff-coloured flanks.
Similar species Both Willow and Marsh Tit are slightly smaller, and have black caps and greyish-black backs.

Range Resident in coniferous forests in northern Eurasia east to the Okhotsk coast and Kamchatka; also in Alaska.
Status in Japan No records.

Tomi Muukkonen

Adult. June, Finland

Azure Tit

Cyanistes cyanus 13cm

Description Small, white and light-blue tit with long tail. White head with dark eye-stripe and nuchal band, light bluish-grey back and white underparts, light-blue wings with white wing-bar and light-blue tail with white corners.

Range Resident in deciduous and mixed forests from western Russia east through southern Siberia to Ussuriland and northern China.
Status in Japan Vagrant. One record in 1987 on Rishiri-tou off the west coast of Hokkaido.

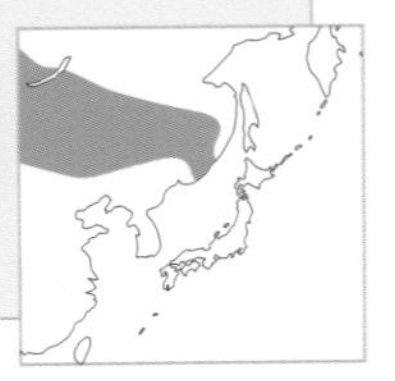

Hanne & Jens Eriksen

Juvenile. July, Tamga, Kyrgyzstan

Great Tit

Parus major 14–15cm

Tadao Shimba

Adult *minor*. January, Shizuoka

Description Black head and throat, white cheeks, yellowish-green back, dark wings with bluish-grey edges and wing-bar. White underparts with black belly stripe, narrower on female. Four races in Japan; southern race is darkest. Race *nigriloris* on the Yaeyama Islands has dark-grey back. **Voice** Most familiar song is a slow, metallic *seeh-wheee* rising on second syllable. Contact call is *tsi-tsi* or *zi-zi*.

Tadao Shimba

Adult *minor*. February, Yamanashi

Range Widespread in temperate Eurasia and a common resident in north-eastern Asia.
Status in Japan Widespread and common resident throughout wooded areas. Forms flocks in winter and often associates with other species in mixed feeding flocks.

Tadao Shimba

Adult *nigriloris*. April, Ishigaki-jima

Norio Yamagata

Adult *nigriloris*. March, Ishigaki-jima

Chinese Penduline Tit

Remiz consobrinus 11cm

Description Small tit which lives in reedbeds. Male has pointed bluish-grey bill, grey head, white face with distinctive broad black mask, chestnut-brown back, dark wings and tail, and pale-brownish belly. Female is browner, has brown head and dark-brown mask.
Voice Thin, metallic *tseeh-tseeh*; often first detected by call.

Akira Yamamoto

Adult ♂ breeding. May, Tsushima

Range Breeds in eastern Asia from Amur Basin to north-eastern China. Winters in Korea, eastern China and Japan.
Status in Japan Considered a vagrant to Kyushu until the 1970s but sightings have since increased. Currently a regular if uncommon winter visitor to central Honshu southwards.

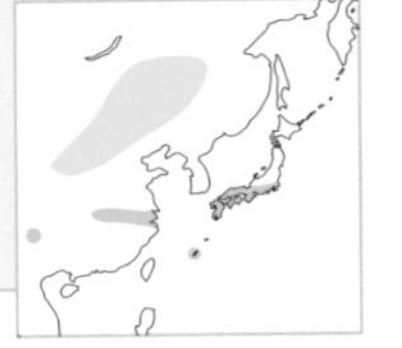

Tadao Shimba

Adult ♂ non-breeding. April, Aichi

Tadao Shimba

Adult ♀ non-breeding. April, Aichi

Eurasian Nuthatch

Sitta europaea 13.5cm

Tadao Shimba

Adult *asiatica*. January, Hokkaido

Description Short-tailed, bluish-grey bird which moves rapidly up, down and around tree trunks. White face with black eye-stripe and white underparts. Three races in Japan. Race *amurensis* (Honshu southwards) has pale-rufous flanks and reddish-brown undertail-coverts. Race *asiatica* on Hokkaido has white flanks.
Voice Song is loud, slow *feeh-feeh-feeh*, contact call is *chwit-chwit*.
Similar species Chinese Nuthatch is smaller, and has buff-coloured underparts.

Tadao Shimba

Adult *asiatica*. January, Hokkaido

Range Widespread in Eurasia and a common resident in north-eastern Asia.
Status in Japan Common resident in deciduous forests from Kyushu northwards. Often joins mixed-species flocks in winter.

Tokio Sugiyama

Adult *amurensis*. February, Nara

Chinese Nuthatch

Sitta villosa 12cm

Description Small nuthatch. Male has a black crown and eye-stripe with prominent white supercilium, dark bluish-grey upperparts and cinnamon-coloured underparts. Female has bluish-grey crown and upperparts and buff-coloured underparts.
Similar species Eurasian Nuthatch is larger with a white breast and upper belly.

Range Uncommon resident in central China and northern Korea. Inhabits coniferous forests.
Status in Japan No records.

Hyun-tae Kim

Adult ♀. November, Hokyong Kwak, South Korea

Eurasian Treecreeper

Certhia familiaris 13.5cm

Description A well-camouflaged white-spotted brown bird with a long, thin downcurved bill, white supercilium and white underparts. Often seen creeping up tree trunks from the base before flying to another tree.
Voice Song is a series of melodious trills; contact call is a thin *tsee*.

Range Widespread in Eurasia; uncommon resident in north-eastern Asia.
Status in Japan Uncommon resident in coniferous forests on Shikoku, Honshu and Hokkaido. Solitary outside breeding season, but sometimes joins mixed-species flocks in winter.

Tadao Shimba

Adult. January, Hokkaido

Japanese White-eye

Zosterops japonicus 12cm

1/16/09 spotted next to our quarters - a pair

Tadao Shimba

Adult *stejnegeri*. June, Miyake-jima

Description Yellowish-green face with pointed dark bill and distinctive white eye-ring. Dark greyish-brown wings and tail. Yellow throat, pale greyish-brown underparts. Several races in Japan; southern races are paler. Race *loochooensis* on the Ryukyu Islands has white belly and flanks.

Voice Song is an erratic series of loud squeaking notes, contact call is a metallic *tseeh-tseeh*.

Similar species Chestnut-flanked White-eye has diagnostic chestnut flank patch.

Tadao Shimba

Adult *japonicus*. February, Shizuoka

Range Japan and southern Korea south to south-eastern China and northern part of south-eastern Asia.

Status in Japan Common resident in wooded areas. Common in city parks and gardens in winter. Often feeds on nectar in hedgerows of *Camellia*.

Yoshihito Goto

Adult *loochooensis*. December, Amami-Oshima

Tadao Shimba

Adult *japonicus*. February, Shizuoka

Chestnut-flanked White-eye

Zosterops erythropleurus 12cm

Description Yellowish-green back, white eye-ring, dark bill with pinkish base to lower mandible. White breast and belly with distinctive chestnut flanks.
Voice Similar to Japanese White-eye, a metallic *tseeh-tseeh.*
Similar species Japanese White-eye has slightly darker back, light greyish-brown underparts and lacks chestnut flank patch.

Han-soo Seo

Adult. April, Chungnam, South Korea

Han-soo Seo

Adult. April, Chungnam, South Korea

Range Breeds in north-eastern China and Ussuriland and winters in south-east Asia.
Status in Japan Rare passage migrant along the coasts of the Sea of Japan.

Teruaki Ishii

Adult. October, Hegura-jima

Bonin Honeyeater

Apalopteron familiare 14cm

Mitsuo Imai

Adult. August, Haha-jima, Ogasawara Islands

Description Olive-green head, back and tail. Yellowish face with distinctive triangle-shaped black eye-patch and long black bill. Yellow underparts.
Voice A plaintive *few woo.*
Similar species Japanese White-eye is smaller and lacks triangular face patch.

Mitsuo Imai

Adult. August, Haha-jima, Ogasawara Islands

Range Endemic to the Ogasawara Islands in Japan.
Status in Japan Inhabits evergreen forests and easily found on Haha-jima. Often feeds on the ground and is approachable.

Mitsuo Imai

Adult. August, Haha-jima, Ogasawara Islands

Black-naped Oriole

Oriolus chinensis 28cm

Description Golden-yellow overall with prominent black eye-stripe and nape band. Also a reddish-orange bill, black primaries and black tail with yellow edge. Female has greenish-yellow back and narrower eye-stripe. Immature is dark greenish-yellow above with dark streaks on belly.
Voice Loud, liquid and flute-like whistle.

Norio Yamagata

Adult ♂. May, Hegura-jima

Range Breeds from south-eastern Transbaikalia east to Amur Basin and Ussuriland, south to south-east Asia. In north-east Asia, a common summer visitor from Korea north to Amur Basin.
Status in Japan Rare passage migrant and mainly recorded in spring. Regularly observed on Hegura-jima in late May. There are a few breeding records.

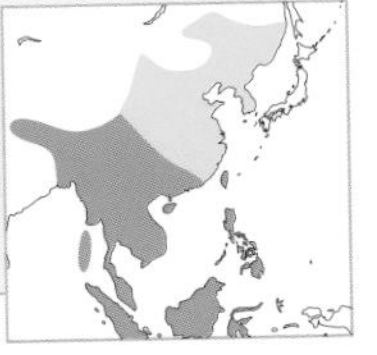

Norio Yamagata

First-summer. May, Hegura-jima

Mitsuo Imai

First-summer. May, Hegura-jima

Tiger Shrike

Lanius tigrinus 17–18cm

Akira Yamamoto

Adult ♀. May, Hegura-jima

Description Male has a distinctive bluish-grey crown and nape with black mask; also a finely barred reddish-brown back and tail and white underparts. Female is duller, and has dark-brown bars on flanks.

Voice Call is a harsh and loud chattering, *jahjahjahjahjahjahjahjah.*

Range Breeds from Ussuriland and north-eastern China south to eastern China. Winters south to south-east Asia.

Status in Japan Rare and local summer visitor to open lowland forests from central Honshu northwards. Has declined sharply since the 1980s.

Tokio Sugiyama

Adult ♂. June, Nagano

Bull-headed Shrike

Lanius bucephalus 19–20cm

Description The most common shrike in Japan. Male has a reddish-brown crown and nape, conspicuous black mask and pale supercilium and throat. Grey back with a white spot at the base of primaries and whitish belly with rusty flanks. Individuals breeding at higher altitudes acquire light-grey back and white underparts. Female and juvenile are browner, have dark brown masks and fine scaly-marked underparts.
Voice Song is a quiet warble interspersed with imitations of other bird calls. Also sharp and loud *cheh-cheh-cheh* calls when establishing winter territory in early autumn.
Similar species Brown Shrike has reddish-brown back.

Range Breeds from southern Ussuriland and north-eastern China south to eastern China. Winters in southern Korea, eastern China and Japan.
Status in Japan Common resident in open forests and cultivated fields. Northern population moves south in winter and common in wooded urban areas and city parks. Rare winter visitor to the Ryukyu Islands.

Tokio Sugiyama

Adult ♂. February, Aichi

Akira Yamamoto

Adult ♀. February, Aichi

Tadao Shimba

Juvenile. September, Mie

Brown Shrike

Lanius cristatus 18–20cm

Tadao Shimba

Adult ♂ *superciliosus*. May, Hegura-jima

Norio Yamagata

First-summer ♂ *superciliosus*. May, Hegura-jima

Yoshihito Goto

Juvenile *superciliosus*. October, Hegura-jima

Description Male *superciliosus has* reddish-brown upperparts with broad white supercilium and forehead; prominent dark eye-stripe and white underparts with pale orange flanks. Female is duller with fine dark speckles on flanks. Male *lucionensis* has grey head, grey-brown back and pale orange-brown underparts; female has dark scaly markings below. Nominate *cristatus* has a darker back.
Voice Loud and harsh chattering *jahjahjahjahjahjah.*
Similar species Isabelline Shrike has sandy-brown back, rusty tail and white spots at the base of primaries.

Range Breeds in eastern Asia from the Altai east to Kamchatka, south to south and west China. Winters in southern Asia from India east to south-east Asia.
Status in Japan Three races recorded; *superciliosus* is an uncommon summer visitor to open forests and grassy fields from Honshu northwards; *lucionensis* is common passage and winter visitor to the Ryukyu Islands, *cristatus* is a rare visitor.

Norio Yamagata

Adult ♂ *lucionensis*. May, Tsushima

Long-tailed Shrike

Lanius schach 24–25cm

Description Large shrike with grey crown and nape, black forehead and broad black mask. Rufous scapulars with black wings and long black tail, whitish underparts with rufous wash on flanks and undertail-coverts.
Voice Loud and harsh chattering.
Similar species Unmistakable due to long black tail.

Kazuyasu Kisaichi

Adult. December, Kanagawa

Range Breeds from western Asia east through the Indian subcontinent to eastern China and Taiwan, south to south-east Asia.
Status in Japan Rare visitor. Most often seen on Yaeyama Islands. Occurs in cultivated and grassy fields.

Great Grey Shrike

Lanius excubitor 24–25cm

Description Grey head and back, black mask, black wings with white patch at the base of primaries. White uppertail-coverts and black tail with white outer-tail feathers. Immature is browner with fine scaly marks on underparts.
Voice Song is a series of quiet warbles with harsh notes interspersed with imitations of other bird calls. Call is a harsh chattering.
Similar species Chinese Grey Shrike is larger, has larger white wing patches and grey upper-tail-coverts.

Yoshio Osawa

Adult. November, Tobishima

Range Breeds in northern Eurasia from Europe east to north-eastern Russia; also in North America. In north-eastern Asia, breeds from south of Arctic Circle to Sakhalin and winters south to northern China. Rare winter visitor further south.
Status in Japan Rare winter visitor to open montane forests and highland meadows on Honshu and Hokkaido.

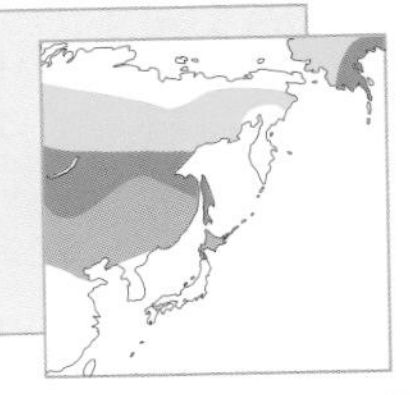

Chinese Grey Shrike

Lanius sphenocercus 30–31cm

Tadao Shimba

Adult. November, Ishikawa

Description Large grey shrike with a prominent black mask, narrow white supercilium and dark wings with large white patches; also a long black tail with white outer tail feathers.
Voice Loud and harsh chattering.
Similar species Great Grey Shrike is smaller with smaller white wing patches and a white rump.

Teruaki Ishii

Adult. January, Shiga

Range Breeds in eastern Asia from Ussuriland and northern China south to central China. Winters south to south-eastern China and Korea.
Status in Japan Rare winter visitor mainly to western Japan. Occurs in cultivated fields and reedbeds.

Teruaki Ishii

Adult. January, Shiga

Isabelline Shrike

Lanius isabellinus 17–19cm

Description Brown head with dark mask, greyish-brown back and rusty uppertail-coverts and tail. Dark wings with white spot at the base of primaries and pale orange-brown underparts.
Similar species See Brown Shrike.

Range Breeds in central Eurasia from western Asia east to Mongolia. Winters from East Africa east to Pakistan and western India.
Status in Japan First recorded on Okinawa-Hontou in 2003 with a few records since. Occurs in cultivated fields and forest edges.

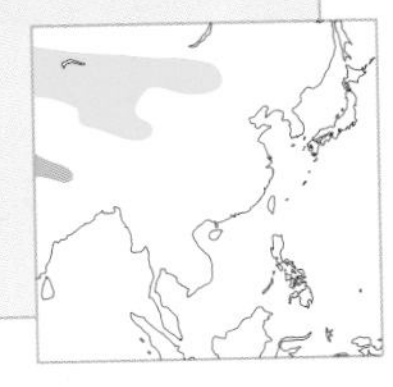

Mitsuo Imai

Adult. December, Okinawa-Hontou

Black Drongo

Dicrurus macrocercus 28cm

Description Long-tailed glossy-black drongo; black bill, glossy purplish-black nape, back and breast; long forked tail and red eye. Immature is browner, has pale undertail-coverts.
Similar species Spangled Drongo is larger, and has a downcurved bill and broad unforked tail. Drongo Cuckoo *Surniculus lugubris,* a vagrant from south-east Asia, is slightly smaller and has distinctive white bars on the undertail-coverts.

Range Breeds from south-eastern Iran east to south-east Asia, Taiwan and southern China, north to northern China. Northern populations are migratory.
Status in Japan Rare visitor to cultivated fields and open woodlands. Often seen on offshore islands in Sea of Japan in spring. Have been a few winter records.

Akira Yamamoto

Adult. May, Tsushima

Spangled Drongo

Dicrurus hottentottus 31–32cm

Kenji Takehara

Adult. October, Okinawa-Hontou

Description Large black drongo with long downcurved bill and broad fan-shaped tail. Glossy purplish-black on crown and breast and glossy greenish-black on wings and tail. Long thin curled hair-like plumes on forehead only visible at close range.
Similar species Black Drongo is smaller, and has deeply forked tail.

Range Breeds in southern Asia from India east to southern China and Philippines, south to South-east Asia. Northern population is migratory.
Status in Japan First recorded in April 2000 on Yonagunijima. Several sightings from Ryukyu Islands since then. Occurs in evergreen forests and cultivated fields.

White-breasted Woodswallow

Artamus leucorhynchus 17.5cm

Tadao Shimba

Adult. December, Queensland, Australia

Description Small swallow-like songbird. Sooty-grey head, back and tail with white underparts and uppertail-coverts. Bluish-grey bill and dark legs. Aerial behavior catching insects in the air.

Range Breeds from Australia and New Guinea north to Philippines.
Status in Japan Vagrant. First recorded in April 1973 on Iriomotejima. Occurs in open forests and cultivated fields. Also recorded once in Korea.

Eurasian Jay

Garrulus glandarius 33cm

Description White crown with black streaks, white eye, bluish-grey bill and black patch at the base of bill. White throat, purplish-brown back and belly and black tail, white undertail-coverts. In flight, conspicuous blue wing patches and white rump. Four races in Japan; *brandtii* on Hokkaido has brown head and dark-brown eyes.
Voice Loud, harsh screaming *kraaah-kraaah*, also imitates other birds.

Kazuyasu Kisaichi

Adult *japonicus*. April, Nagano

Range Widespread across temperate zone of Eurasia from western Europe east to Ussuriland and Sakhalin, south to southern China and Taiwan. Common resident in north-eastern China and Korea.
Status in Japan Common resident in mountain forests from Kyushu northwards. Moves to lower altitudes in winter.

Kazuyasu Kisaichi

Adult *japonicus*. April, Nagano

Mitsuo Imai

Adult *brandtii*. March, Hokkaido

Lidth's Jay

Garrulus lidthi 38cm

Kazuyasu Kisaichi

Adult. March, Amami-Oshima

Description Dark-blue head, breast and neck with pale bill with black patch at base. Chestnut-black belly and dark blue wings and tail, dark legs. In flight, white trailing edge to primaries and white tips to tail. Juvenile is browner.
Voice Loud, harsh screaming *arrrrrr-arrrrrr*.

Yoshihito Goto

Adult. September, Amami-Oshima

Range Endemic to Amami Islands in Japan.
Status in Japan Uncommon resident in evergreen forests and cultivated fields. Often breeds near human habitation.

Tokio Sugiyama

Adult. March, Amami-Oshima

Spotted Nutcracker

Nucifraga caryocatactes 34–35cm

Description Small crow that inhabits mountain forests. Dark-brown head, back, and belly with white streaks. Dark purplish-brown wings and tail with white tips, white undertail-coverts.
Voice Loud, hoarse and repeated *kra-kra.*

Range Breeds across northern Eurasia from Alps east to Okhotsk coast, Kamchatka Peninsula and Sakhalin, south through north-eastern China to central China and Taiwan. Rare resident in Korea.
Status in Japan Uncommon resident in high mountains on Hokkaido, Honshu and Shikoku. Moves to lower altitudes in winter.

Tadao Shimba

Adult. July, Yamanashi

Tadao Shimba

Adult. July, Yamanashi

Tadao Shimba

Adult. July, Yamanashi

Common Magpie

Pica pica 45cm

Tadao Shimba

Adult. February, Fukuoka

Description Black and white bird with long tail. Black head, back and breast. White belly and black undertail-coverts. Glossy greenish-black wings with white primaries, white shoulders and glossy-greenish black tail.

Voice Call is a series of harsh, metallic *chak* notes.

Kazuyasu Kisaichi

Adult. March, Saga

Range Widespread in temperate zone of Eurasia from Europe east to Ussuriland and North-eastern China, south to southern China and Taiwan, also in northern Africa and western North America. Common resident in Korea.

Status in Japan Locally common resident in northern Kyushu, rare elsewhere. Inhabits cultivated fields and wooded urban areas. Nests in trees and on telephone poles.

Tim Edelsten

Adult. February, Seolleung, South Korea

Azure-winged Magpie

Cyanopica cyana 35–37cm

Description Medium-sized bird with long tail. Black cap, dark bill, white throat and greyish back and belly. In flight, light-blue forewings and secondaries contrast with black primaries and white tip to tail. Juvenile has white spots on black cap and browner back.
Voice Hoarse, longish *zcheueee,* often followed by a short *kweet-kweet.*

Manabu Totsuka

Adult. June, Nagano

Range Breeds from northern Mongolia east to Ussuriland and north-eastern China, south to eastern China, also on Iberian Peninsula. Common resident in Korea.
Status in Japan Common resident in foothills and wooded urban areas from central to northern Honshu. Common in parks and gardens in Tokyo. Forms flocks for most of the year.

Norio Yamagata

Adult. December, Ibaragi

Kazuyasu Kisaichi

Adults. March, Chiba

Red-billed Chough

Pyrrhocorax pyrrhocorax 40cm

Valery N. Moseykin

Adult. Altai Mountains, Russia

Description Glossy-black body with long broad wings and short, square tail. Long downcurved red bill and red legs. Juvenile has orange bill.

Voice Loud *klyaw* or *chaw*.

Range Inhabits rocky mountain meadows and steppes in inland Eurasia east to north-eastern China, also on rocky cliffs in coastal Europe. Vagrant to Korea.

Status in Japan No records.

Siberian Jay

Perisoreus infaustus 28cm

Tomi Muukkonen

Adult. August, Finland

Description Dark head with small black bill, greyish-brown back and underparts. In flight, dark-brown wings with rusty-orange wing-bars, rump and tail with dark central feathers.

Voice Mostly silent, but has many different calls, including harsh screams and mewing.

Range Common resident in coniferous forests in northern Eurasia from Scandinavia east to the Okhotsk coast, south to Sakhalin, Ussuriland and Heilongjiang.

Status in Japan No records.

Common Raven

Corvus corax 61cm

Description Large crow with long wings and long wedge-shaped tail. Glossy-black overall with large bill and shaggy throat feathers.
Voice Loud, hoarse *kraa*.
Similar species Large-billed Crow is smaller with peaked forehead and thicker bill.

Yasuko Tashiro

Adult. March, Hokkaido

Range Widely distributed in Eurasia. Breeds from Europe east to Chukotki and Kamchatka Peninsulas, south to Amur Basin, northern China, Sakhalin and Kurile Islands, also in northern Africa and North America. Found in a variety of habitats.
Status in Japan Uncommon winter visitor to eastern Hokkaido. Mainly inhabits rocky cliffs and coastal areas. Recent increase in records from inland areas.

Yasuko Tashiro

Adult. March, Hokkaido

Manabu Totsuka

Adult. March, Hokkaido

Carrion Crow

Corvus corone 50cm

Tadao Shimba

Adult. April, Aichi

Tadao Shimba

Adult. February, Nagano

Description Overall glossy-black. Smooth round head, black bill and legs.
Voice Loud, hoarse *kraaa*.
Similar species Large-billed Crow is slightly larger and has peaked forehead, much thicker bill and different call. Rook has thinner pointed bill.

Range Widespread in Eurasia from Europe east to Okhotsk coast, Ussuriland and Sakhalin, south to northern China, Korea and Japan.
Status in Japan Abundant resident from Kyushu northwards. Cultivated fields, coastal areas and wooded urban areas. Forms large flocks in winter.

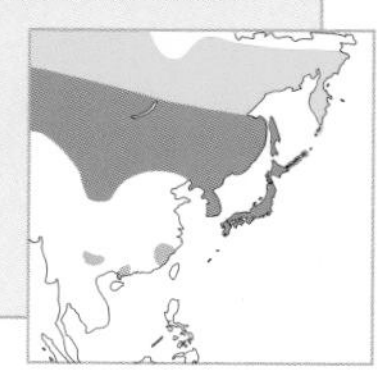

Tadao Shimba

Adults. December, Hokkaido

Large-billed Crow

Corvus macrorhynchos 56.5cm

Description Overall glossy-black with peaked forehead and black legs. Thick black bill with down-curved upper mandible. Four races in Japan; race *osai* on Yaeyama Islands is the smallest and has a rounder head.
Voice Loud, clear *krroh* or *kaw*.
Similar species Carrion Crow is slightly smaller, has rounder head and straight black bill.

Range Breeds from western Asia east to eastern China and South-east Asia, north to Amur Basin, Ussuriland, Sakhalin and north-eastern China. Common resident in Korea.
Status in Japan Abundant resident in forests, cultivated fields and wooded urban areas. Inhabits centres of large cities in eastern Japan.

Tadao Shimba

Juvenile *japonensis*. July, Tokyo

Tadao Shimba

Adult *japonensis*. September, Aichi

Tadao Shimba

Adult *osai*. April, Ishigaki-jima

Rook

Corvus frugilegus 47cm

Tadao Shimba

Adult. January, Kagoshima

Description Glossy-black overall. Thin, pointed, dark bill with pale base. Immature is browner and has entirely dark bill.
Voice Loud, hoarse *graaa*.
Similar species Carrion Crow has thicker bill.

Range Breeds across Eurasia from Europe east to Amur Basin and Ussuriland, south through north-eastern China to eastern China. Moves south in winter and common winter visitor to Korea.
Status in Japan Locally common winter visitor to cultivated fields in Kyushu. Forms large flocks. Wintering range expanding to east and north in recent years.

Western Jackdaw

Corvus monedula 30–34cm

Kazuyasu Kisaichi

Adult. December, Hokkaido

Description Small crow with short, dark bill and white eyes. Black crown, face and throat. Greyish ear-coverts, black head with greyish-white patch on hindneck. Sooty-black belly and black back. Immature is browner and has dark eyes.
Voice Metallic *kyak,* identical to Daurian Jackdaw.
Similar species See Daurian Jackdaw.

Range Widespread in western and central Eurasia east to about Transbaikalia. Northern population is migratory.
Status in Japan
Vagrant. Two records from Hokkaido. First recorded on Teuri-tou off west coast of Hokkaido in 1986.

Daurian Jackdaw

Corvus dauuricus 33cm

Description Small crow with short, dark bill and dark eyes. Two distinct colour morphs; dark-morph has black head, face and belly, sooty-black back and greyish streaks on ear-coverts; light-morph has distinctive white collar and belly with sooty-black back and white streaks on ear-coverts.
Voice A cutting, metallic *kyak* or *chak*.
Similar species Western Jackdaw has white eyes and greyish-white patch on hindneck. Immature Eurasian Jackdaw has dark eyes and can be difficult to separate.

Tadao Shimba

Adult dark phase. January, Kagoshima

Range Breeds from Mongolia east to Amur basin, Ussuriland and North-eastern China, south to central China. Winters south to South-eastern China and uncommon visitor to Korea.
Status in Japan Rare winter visitor to cultivated fields in Kyushu. Often associated with flocks of Rooks. Dark-morph birds more commonly observed. Sightings have increased in recent years and wintering range expanding to east and north.

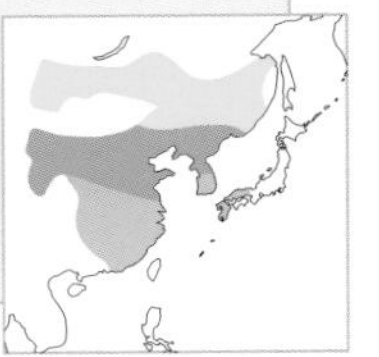

Tadao Shimba

Juvenile dark phase. February, Aichi

Tadao Shimba

Adult light phase. January, Kagoshima

Red-billed Starling

Sturnus sericeus 24cm

Tokio Sugiyama

Adult ♂. January, Kagoshima

Description Male has pale-yellowish hood with dark border, purplish-grey underparts, black primaries with white at base and pale rump. Also a purplish-grey belly, white undertail-coverts and black tail. Dark eyes; dark-tipped, pointed red bill and orange legs. Female has browner head, back and belly.

Similar species White-shouldered Starling is paler, has distinctive white patch on upperwing-coverts.

Tokio Sugiyama

Adult ♀. January, Kagoshima

Range Breeds in south-eastern China. Some move south in winter.

Status in Japan Rare winter visitor to cultivated fields and wooded urban areas in southern Japan. Winters regularly on the Yaeyama Islands and often seen among flocks of White-shouldered Starlings.

Mitsuo Imai

Adult ♂. May, Hegura-jima

Daurian Starling

Sturnus sturninus 17–18cm

Description Male has light-greyish head, breast and belly. Also a dark-brown patch on rear crown, black back and glossy greenish-black wings with white edges to upperwing-coverts. Thick, dark bill, dark eyes and legs. Female has greyish-brown head and dark-brown back.
Similar species Female Chestnut-cheeked Starling has a browner back and no wing-bars.

Range Breeds from Transbaikalia east to the Amur Basin and Ussuriland, south to north-eastern China and northern Korea. Winters in southern China and south-east Asia.
Status in Japan Rare passage migrant to offshore islands in the Sea of Japan. Mostly recorded on spring migration and often seen on Hegura-jima in late May.

Mitsuo Imai

Adult ♂. May, Hegura-jima

Norio Yamagata

Adult ♂. May, Hegura-jima

Chestnut-cheeked Starling

Sturnus philippensis 18–19cm

Norio Yamagata

Adult ♂. April, Aichi

Kazuyasu Kisaichi

Juvenile. July, Ibaragi

Description Male has cream-coloured head with chestnut cheeks and ear-coverts; also glossy purplish-black above with white upper median-coverts, pale rump and pale undertail-coverts. Dark eyes and bill and dark legs. Female has greyish-brown head and belly, dark-brown back and black wings and tail. Juvenile is similar to female, but paler.
Voice Song is a series of loud melodious squeaks, call is a harsh, loud squeak.
Similar species Female Daurian Starling has darker back, white shoulder patch and thinner bill.

Range Breeds in southern Sakhalin and Japan; winters mainly in Philippines and Borneo.
Status in Japan Common summer visitor to central Honshu northwards, uncommon passage migrant elsewhere. Breeds in tree hollows in montane deciduous forests.

Tokio Sugiyama

Adult ♂. May, Nagano

White-shouldered Starling

Sturnus sinensis 19cm

Description A light brownish-grey starling with white eyes and bluish-grey bill and legs. Male has a conspicuous white shoulder-patch contrasting with black primaries. Female is browner with smaller white shoulder-patch.
Similar species Daurian Starling has black bill, dark brown back and dark eyes.

Range Breeds in southern China and northern south-east Asia; winters in south-east Asia and Taiwan.
Status in Japan Rare winter visitor to cultivated fields and wooded urban areas in southern Japan. Small flocks regularly winter on the Ryukyu Islands.

Tadao Shimba

Adult ♂. April, Mie

Tokio Sugiyama

Adult ♂. January, Kagoshima

Tokio Sugiyama

Adult ♂. January, Kagoshima

Rosy Starling

Sturnus roseus 21cm

Mitsuo Imai

First-winter. January, Okinawa-Hontou

Description Adult is unmistakable. Purplish-black head, breast and nape contrast with pinkish-brown back and underparts. Black wings, tail and undertail-coverts. Juvenile has yellowish bill, pale greyish-brown upperparts, dark-brown wings and tail, buff-coloured underparts and dark-brownish undertail-coverts with buff-coloured edges.

Range Breeds in Eurasia from the Balkans east through central Asia to north-western China. Winters mainly in Indian subcontinent.

Status in Japan Very rare winter visitor and passage migrant. Recorded on Hegura-jima and the Ryukyu Islands.

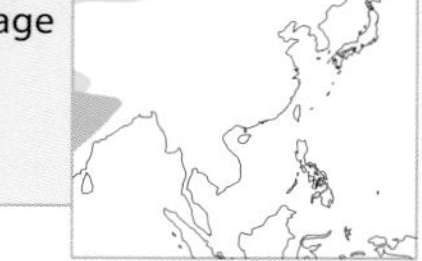

Common Starling

Sturnus vulgaris 21–23cm

Akira Yamamoto

Adult non-breeding. November, Aichi

Description In breeding plumage, overall glossy-black with yellow bill, dark-red legs and dark-brown wings with buff-coloured edges. In non-breeding plumage, dark bill and white spots all over body. Juvenile is browner.

Range Widely distributed in Eurasia east to Transbaikalia; also introduced to the Americas and Australia. Forms large flocks and moves south in winter.

Status in Japan Rare winter visitor to cultivated fields and wooded urban areas. First recorded in 1969 and now winters regularly at a few places in western Japan. Sightings have increased in recent years.

White-cheeked Starling

Sturnus cineraceus 24cm

Description Male has a dark head with white forehead and cheeks, orange-yellow bill with dark tip and yellow legs. Otherwise generally dark brown with white uppertail- and undertail-coverts and a white-tipped tail. Female and juvenile are browner.

Range Breeds from eastern Mongolia east to Sakhalin, Ussuriland and northern China, south to Korea and Japan. Winters in southern China, Korea and Japan.
Status in Japan The most common starling in Japan. Abundant resident in cultivated fields and wooded urban areas. Often nests in house attics or in cracks in walls. Winter visitor to the Ryukyu Islands. Forms large flocks in winter.

Tokio Sugiyama

Adult ♂. November, Aichi

Tadao Shimba

Adult ♀. April, Aichi

Akira Yamamoto

Juvenile. June, Aichi

House Sparrow

Passer domesticus 17cm

Mitsuo Imai

Adult ♀. May, Hegura-jima

Description Male has reddish-brown head with grey crown, black lores and black bib. Also greyish-white cheeks and belly, reddish-brown back with black streaks and white edges to median-coverts. Female has greyish-brown head with buff-coloured supercilium behind eye and pale greyish-brown underparts.
Voice Song is a simple series of chirps. Call is a short, monotonous *chirrup* or *chirp.*
Similar species Eurasian Tree Sparrow has distinctive black cheek spot. Female Russet Sparrow has reddish-brown back and head with white supercilium behind eye.

Range Widespread in Eurasia, east to western China and Amurland and in Africa; introduced to Americas and Australia. Mostly absent from north-east Asia.
Status in Japan Vagrant. First recorded in February 1990 on Rishiri-tou off northern coast of Hokkaido. A few records since then and has interbred with Eurasian Tree Sparrow in Hokkaido.

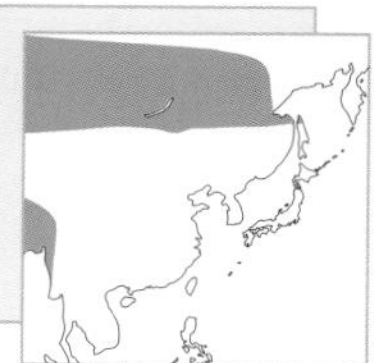

Eurasian Tree Sparrow

Passer montanus 14–15cm

Tadao Shimba

Adult. February, Nagano

Description Both sexes have a brown head, white cheeks, black bib and stout black bill. Distinctive black spot on cheeks and white collar. Brown back with black streaks, two wing-bars and whitish underparts with brownish flanks; legs pinkish. Juvenile is paler, with smaller black cheek spot.
Voice Short, metallic *chik* or *tek.*
Similar species Male Russet Sparrow has reddish-brown head and lacks black cheek spot. Male House Sparrow has grey crown, larger bib and lacks black cheek spot.

Range Widespread throughout Eurasia except northern Siberia and the Kamchatka Peninsula.
Status in Japan Abundant resident in wooded urban areas and cultivated fields. Forms large flocks in winter.

Russet Sparrow

Passer rutilans 14cm

Description Male has reddish-brown head, black lores and bib, and white cheeks. White underparts and a reddish-brown back with black streaks. Wings black with white wing-bar. Acquires white supercilium in non-breeding plumage. Female has greyish-brown head with broad white supercilium behind eye and buff-coloured belly.
Voice Similar to Eurasian Tree Sparrow, but softer and thinner *cheeep* or *chilip*.
Similar species Female House Sparrow is greyer, has buff-coloured supercilium and greyish rump.

Tadao Shimba

Adult ♂ breeding. April, Aichi

Range Breeds in southern Asia from the Himalayas east to eastern China and Taiwan, north to Japan and southern Sakhalin. Northern population is migratory.
Status in Japan Breeds locally in broad-leaved deciduous montane forests from central Honshu northwards. Winters in southern Japan and is locally common in rice fields and cultivated fields. Forms large flocks in winter.

Tadao Shimba

Adult ♂ breeding. April, Aichi

Tadao Shimba

Adult ♀. April, Aichi

Brambling

Fringilla montifringilla 16cm

Tadao Shimba

Adult ♂ breeding. May, Hegura-jima

Description Male in breeding plumage has black head and back with diagnostic orange breast and shoulders; black wings with orange-fringed wing-coverts and tertials, conspicuous white rump, white belly and dark spotted flanks. Head becomes mottled brown and black in non-breeding plumage. Female is duller, has black stripe at sides of head and grey nape. **Voice** Song is a hoarse, monotonous *dreeee* and call is a harsh *chet*.

Mitsuo Imai

Adult ♂ breeding (in moult). May, Hegura-jima

Range Breeds across northern Eurasia from Scandinavia east to the Okhotsk coast and Kamchatka Peninsula, south to Sakhalin. Winters south to Korea, Japan and southern China. **Status in Japan** Common winter visitor to wooded areas and cultivated fields. Often visits bird feeders in flocks mixed with Oriental Greenfinch.

Tokio Sugiyama

Adult ♀♀. January, Shiga

Asian Rosy Finch

Leucosticte arctoa 16cm

Description Male has black forehead, face and throat, yellow bill and a buff-coloured patch on rear crown and nape. Black above with buff-coloured feather edges, dark-brown wings with purplish-red wing coverts. Black below with purplish-red spots and forked, dark-brown tail. Female is duller.
Voice Song includes a series of *chew* notes and call is a single, whistled *pert* or *chew*.

Akira Yamamoto

Adult ♂. December, Nagano

Range Breeds in central and southern Siberia from the Altai east to northern Amurland and Kamchatka, south to north-western China. Winters at lower elevations and in warmer areas.
Status in Japan Uncommon winter visitor to rocky cliffs, stony coastal areas and cultivated fields. A few may breed in mountains on Hokkaido.

Akira Yamamoto

Adult ♀. December, Nagano

Akira Yamamoto

Adult ♂. December, Nagano

Pine Grosbeak

Pinicola enucleator 20cm

Manabu Totsuka

Adult ♂. June, Hokkaido

Teruaki Ishi

Juvenile ♀. November, Hegura-jima

Description Large, plump finch with a long tail and stout, rounded bill. Male has deep pinkish-red head, back and breast, dark-brown wings with two white wing-bars and white edges to tertials, greyish flanks, lower belly and undertail-coverts and forked dark-brown tail. Female has olive-yellow head, back and breast.
Voice Song is a loud melodious warble. Call is a fluty whistled *pui pui pui*.

Range Breeds in the taiga of northern Eurasia, from Scandinavia east to Kamchatka, south to Amur Basin, Sakhalin and northern Japan; also in North America. Sedentary or partially migratory. Moves erratically to regions south of breeding range when food is scarce.
Status in Japan Uncommon resident in Hokkaido. Breeds in coniferous forests in mountains, moving to lower elevations in winter, rare elsewhere. Occasionally visits roadsides and gardens in cities searching for berries; usually tame and approachable.

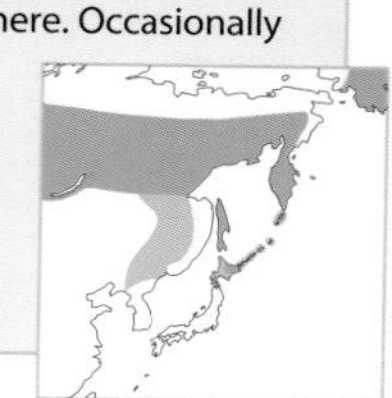

Teruaki Ishi

Adult ♂. March, Kyoto

Pallas's Rosefinch

Carpodacus roseus 16-17cm

Description Male has rosy-red head and underparts with silvery-white forehead and throat. Rosy back with red streaks, dark-brown wings with pinkish-white edges to wing-coverts and white edges to tertials, dark-brown forked tail. Female is pale-brown overall with dark–streaked back, breast and flanks. Immature male is similar to female but has pale reddish head and breast.
Voice Call is a short and soft whistled *zit*.
Similar species Female Common Rosefinch is smaller, browner and has less streaked back.

Tokio Sugiyama

Adult ♂. February, Aichi

Range Breeds in central and eastern Siberia from the Altai and Yenisey east to the Okhotsk coast and Sakhalin, south to northern Mongolia and Amurland. Winters south to Ussuriland, northern China and Korea.
Status in Japan Uncommon winter visitor to mountain forests and forest edges with dense thickets in northern Japan. Feeds in small flocks on seeds and berries. Sometimes visits bird feeders when food is scarce.

Akira Yamamoto

First-winter ♂. February, Aichi

Tadao Shimba

First-winter ♀. February, Aichi

Common Rosefinch

Carpodacus erythrinus 14cm

Tomi Muukkonen

Adult ♂. May, Mustasaari, Finland

Description Male has deep-red head, breast and back, dark-brown wings and tail and whitish underparts. Female and juvenile are uniform olive-brown above with indistinct dark streaks, buff-coloured edges to greater and median-coverts and whitish underparts.
Similar species Female Pallas's Rosefinch is larger and has heavily streaked back and underparts.

Range Breeds across Eurasia from North-eastern Europe east through Siberia to Kamchatka, south to the Amur Basin, Ussuriland, northern Sakhalin and north-eastern China; also in Himalayas. Winters south to southern Asia.
Status in Japan Rare passage migrant seen on forest edges and cultivated fields with thickets.

Arctic Redpoll

Carduelis hornemanni 13cm

Norio Yamagata

Adult. March, Hokkaido

Description Similar to Common Redpoll, but whiter. Male has greyish-white head with small red patch on forehead and tiny triangular bill. Greyish-brown back with dark streaks, unstreaked white rump and dark-brown forked tail; underparts whitish with streaked flanks. Breast is usually paler than in Common Redpoll.
Voice Similar to Common Redpoll.

Range Breeds on arctic tundra in Eurasia and North America. Sedentary and partially migratory. In north-eastern Asia, occasionally winters south to Korea.
Status in Japan Very rare winter visitor to northern Japan. Often seen in flocks of Common Redpoll.

Common Redpoll

Carduelis flammea 13.5cm

Description Small greyish-white finch with small yellow bill. Male has a red forehead, greyish-white back with greyish-brown streaks and white rump with fine dark streaks. Also dark-brown wings with white edges to wing-coverts, dark-brown forked tail and white underparts with pinkish breast and brown streaks on flanks. Female browner and has a whiter breast.
Voice Song is a variety of trills. Call is a harsh and metallic *dyuee*.
Similar species Arctic Redpoll has tiny bill, smaller red forehead patch and pure white rump.

Range Breeds across northern Eurasia from Scandinavia east to the Chukotski and Kamchatka Peninsulas, south to Sakhalin; also in North America. Winters south to temperate regions.
Status in Japan Uncommon winter visitor to northern Japan. Numbers vary from year to year. Inhabits coniferous forests, forest edges and riverbanks.

Norio Yamagata

Adult ♂. March, Hokkaido

Kazuyasu Kisaichi

First-winter. January, Hokkaido

Shigeo Ozawa

First-winter. March, Hokkaido

Eurasian Siskin

Carduelis spinus 12.5cm

Mitsuo Imai

Adult ♂ breeding. May, Hegura-jima

Mitsuo Imai

Adult ♀ May, Hegura-jima

Description Male has bright-yellow face and head, black forehead and crown, yellowish-green above with conspicuous yellow fringes to greater and median-coverts; yellow rump and forked tail with yellow base to outer-tail feathers. Female is duller, with finely streaked greenish-brown head and back and heavily-streaked whitish underparts. Broad yellow wing-bars contrasts with black wings in flight.
Voice Song is a variety of trilling and twittering; call is a metallic and harsh *toolee.*
Similar species See Common Redpoll.

Range Europe and eastern Asia. In north-east Asia, breeds from Ussuriland south to Sakhalin and northern Japan. Winters south to southern Japan.
Status in Japan Common winter visitor to coniferous and mixed forests. A few breed in northern Japan.

Akira Yamamoto

Adult ♂ non-breeding. October, Hegura-jima

Oriental Greenfinch

Carduelis sinica 14.5cm

Description Male *minor* has a deep green face, dark-greyish crown and hindneck, olive-brown back and black wings with white fringes to tertials and a yellow primary flash. Also a black tail with yellow base to outer-tail feathers, olive-brown underparts and yellow undertail-coverts. Female is browner, with brown face and underparts. Race *kawarahiba* is slightly larger with broader white tertial fringes. Race *kittlitzi* is greener.
Voice Song is a melodious trilling *kirr-kirr korr-korr djeeeen.* Call is a metallic, harsh *cheeeh.*

Range Breeds in eastern Asia from Kamchatka and Sakhalin south through Ussuriland and Korea to southern China. Northern population is migratory.
Status in Japan Race *minor* is a common resident in wooded areas, city parks and gardens; *kawarahiba* is a common winter visitor. Critically endangered race *kittlitzi* occurs on Ogasawara Islands.

Tadao Shimba

Adult ♂ *kawarahiba*. March, Nagano

Tadao Shimba

Adult ♀ *kawarahiba*. February, Aichi

Tadao Shimba

Adult ♂ *minor*. June, Aichi

Common Crossbill

Loxia curvirostra 16.5cm

Tokio Sugiyama

Adult ♂. February, Osaka

Description A stocky, short-tailed, large-billed bird with unique cross-tipped bill. Male is dark red with black wings and deeply forked black tail. Female has olive-green crown and back, greyish-brown face, yellowish rump and pale yellowish underparts. Immature is greyish-brown with dark black streaks. **Voice** Loud and rapidly repeated *glip-glip-glip.* **Similar species** Two-barred Crossbill is smaller and has two white wing–bars.

Tokio Sugiyama

Adult ♀. February, Osaka

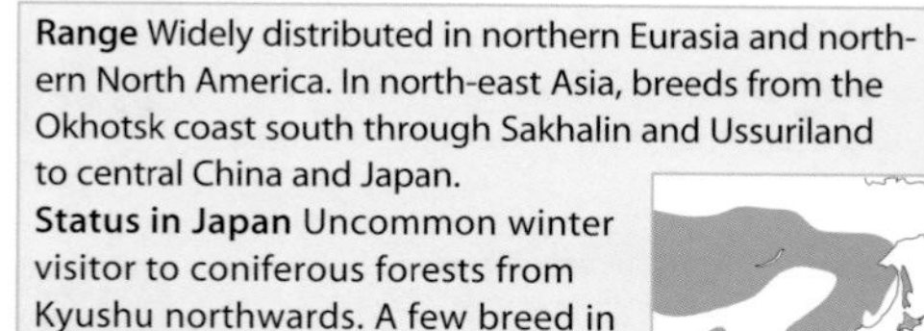

Range Widely distributed in northern Eurasia and northern North America. In north-east Asia, breeds from the Okhotsk coast south through Sakhalin and Ussuriland to central China and Japan.

Status in Japan Uncommon winter visitor to coniferous forests from Kyushu northwards. A few breed in mountain forests from central Honshu northwards.

Two-barred Crossbill

Loxia leucoptera 15cm

René Pop

Adult ♂. October, Holland

Description Similar to Common Crossbill, but slightly smaller and slimmer. Both sexes show prominent white tips to greater and median coverts, forming a distinctive double wingbar. Male is pinkish-red with black wings and forked black tail. Female has olive-green back with yellowish rump and greenish-yellow underparts. Immature is like female but more heavily streaked. **Voice** Call is a rattling *glip-glip* softer than Common Crossbill; also a twittering *chet-chet*.

Range Breeds in taiga in northern Eurasia from northeast Europe east to eastern Siberia and northern China; also in North America. Mainly sedentary, but nomadic in some years.

Status in Japan Rare winter visitor to coniferous forests in northern Japan. Often in mixed flocks with Common Crossbill.

Japanese Grosbeak

Eophona personata 23cm

Description Large finch with large yellow bill. Sexes alike. Glossy purplish-black crown and face, brownish-grey back, purplish-black wing-coverts and black primaries with diagnostic white patch at base. Grey belly, pale-brown flanks and white undertail-coverts, black tail and pinkish legs. Juvenile has greyish-brown head and lacks black crown.
Voice Song is a short series of loud, melodious whistles. Call is a short, loud *tak*.
Similar species Chinese Grosbeak is slightly smaller with orange flanks and white trailing edges to wings; male has glossy purplish-black hood.

Range Breeds in eastern Asia from eastern Amurland and Ussuriland south to northern Korea and Japan. Winters in eastern China and Japan.
Status in Japan Common resident in broad-leaved deciduous forests from Kyushu northwards. Moves south in winter and often seen in city parks and gardens.

Tadao Shimba

Adult. March, Aichi

Tadao Shimba

Adult. November, Aichi

Tadao Shimba

Adults. March, Aichi

Chinese Grosbeak

Eophona migratoria 18.5cm

Tokio Sugiyama

Adult ♂ non-breeding. February, Aichi

Description Medium-large finch with large yellow bill. Male has glossy purplish-black hood, greyish-brown nape, back and belly with rufous flank patch, pale rump and black tail. In flight, black forewings and primaries contrast with dark-blue secondaries and tertials; white trailing edge to wings and conspicuous white patches on primary-coverts. Female has greyish-brown head.
Voice Song is a variety of whistles. Call is a loud, short *tak*.
Similar species See Japanese Grosbeak.

Tadao Shimba

Adult ♀ non-breeding. February, Aichi

Range Breeds in eastern Asia from north-eastern Mongolia east to southern Amurland and Ussuriland, south to northern Korea; also in eastern China. Winters south to southern China.
Status in Japan Uncommon passage migrant and winter visitor to woodlands in western Japan. Numbers fluctuate from year to year. There are a few breeding records from southern Japan.

Tadao Shimba

Adult ♂ non-breeding. February, Aichi

Eurasian Bullfinch

Pyrrhula pyrrhula 15.5cm

Description Male *griseiventris* has black cap, orange-red cheeks and small black bill. Dark-grey back and black wings with greyish-white patch on greater-coverts. White rump, black tail and grey belly, white undertail-coverts. Female has greyish-brown cheeks, back and belly. Male *rosacea* has pale orange-red belly; purplish-brown in female. Male *cassinii* has reddish-orange breast and belly and female is buff below; both sexes have a white wing-patch.
Voice Song is a quiet warble and call is a soft piping *pew*, repeated at intervals.

Range Breeds across northern Eurasia from northern Europe east to Kamchatka and the Okhotsk coast, south to Amur Basin, Ussuriland, Sakhalin and northern Japan. Winters south to temperate regions, uncommon winter visitor to Korea and north-eastern China.
Status in Japan Breeds in coniferous forests in high mountains from central Honshu northwards. Moves to lower elevations in winter. Three races recorded; *griseiventris* breeds in Japan; *rosacea* is an uncommon winter visitor to wooded areas; *casinii* is a rare winter visitor.

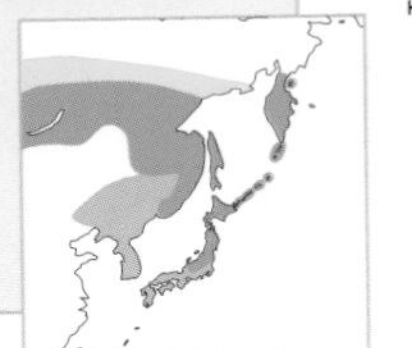

Tokio Sugiyama

Adult ♂ *griseiventris*. May, Yamanashi

Tadao Shimba

Adult ♂ *rosacea*. March, Aichi

Tadao Shimba

Adult ♀ *rosacea*. February, Shiga

Norio Yamagata

Adult ♂ *cassinii*. January, Hokkaido

Norio Yamagata

Adult ♀ *cassinii*. January, Hokkaido

Hawfinch

Coccothraustes coccothraustes 18cm

Kazuyasu Kisaichi

Adult ♂ non-breeding. April, Nagano

Description Medium-large plump finch with a large bill and short white-tipped tail. Male has an orange-brown head, black lores and chin, dark-brown back, grey nape and pale-brown bill; bill becomes bluish-grey in summer. In flight, dark-brown forewings contrast with dark-blue flight feathers and conspicuous white bars on wing-coverts and primaries. Female is similar to male but paler.
Voice Call is a sharp, metallic *tik* or shrill *tzeee.*
Similar species Female Chinese Grosbeak has a longer tail and orange flanks.

Mitsuo Imai

Adult ♀ breeding. May, Hegura-jima

Range Breeds in northern Eurasia from Europe east through southern Siberia to Amurland, Ussuriland and Sakhalin, south to northern Korea and northern Japan; also in north-west Africa. Northern populations move south in winter.
Status in Japan Common winter visitor to open forests, parks and gardens. A few breed in Hokkaido.

Tadao Shimba

Adult ♀ non-breeding. February, Shizuoka

Long-tailed Rosefinch

Uragus sibiricus 15cm

Description Small finch with small, pale bill and long tail. Male has scarlet face, back and belly; white streaks on crown, ear-coverts and throat, dark wings with two white wing-bars and a black tail with white edges. Becomes duller overall in non-breeding plumage. Female has brown head, back and belly with dark streaks.
Voice Song is a melodious variety of trills. Call is a sharp, metallic *pit*.
Similar species Pallas's Rosefinch is slightly larger, has shorter tail and lacks white outer tail feathers.

Range Breeds in southern Siberia from Altai east to Amurland, Ussuriland and Sakhalin, south to northern Korea, also in central China. Northern populations are migratory and winter south to southern Korea and southern Japan.
Status in Japan Locally common summer visitor to coastal thickets and grassy meadows in Hokkaido and northern Honshu. Migrates to Honshu and southwards in winter. In winter, frequently seen in open forests and thickets and along riverbanks.

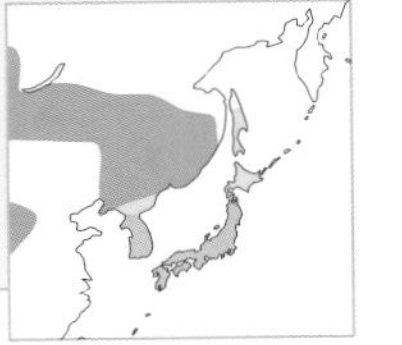

Tadao Shimba

Adult ♂ breeding. June, Hokkaido

Tokio Sugiyama

Adult ♂ non-breeding. March, Aichi

Tokio Sugiyama

Adult ♀. March, Aichi

Yellowhammer

Emberiza citrinella 16.5cm

Tokio Sugiyama

Adult ♂ non-breeding. December, Ohita

Description Male has distinctive yellow head with few dark streaks, brown back with dark streaks and rusty-brown rump. Also a yellowish breast and belly with rusty-brown streaks on breast and flanks. Female and non-breeding male are duller with more heavily streaked breast and belly.
Voice Song is a series of trilling notes followed by longer drawn-out note *tzi-tzi-tzi-tzi-tzi-tzi tzeeeh* and call is a short, sharp *tik*.

Range Breeds in western Eurasia from Europe east to the Altai and Vilyuy River. Partly migratory and winters south to southern Europe and central Asia. Widespread throughout Europe.
Status in Japan Vagrant. Occurs in cultivated fields and forest edges.

Jankowski's Bunting

Emberiza jankowskii 16cm

Göran Ekström

Adult. North-east China

Description Adult male similar to Meadow Bunting, but has grey ear-coverts, whitish throat, paler belly and a diagnostic blackish patch on the centre of belly; paler above with more heavily streaked mantle. In non-breeding plumage, the belly patch is often paler and smaller. Female is also similar to Meadow Bunting, but is paler overall, has more heavily streaked mantle and whiter underparts with streaks on upper breast.

Range Breeds in Jilin province in north-eastern China and winters in central China. Vagrant to Korea. Inhabits dry overgrown sand dunes, with low bushes or trees and open grassy areas. Endangered.
Status in Japan No records.

Pine Bunting

Emberiza leucocephalos 17cm

Description Breeding male has distinctive chestnut, black and white head pattern, reddish-brown back with dark streaks and whitish under-parts with chestnut streaks on breast and flanks. Winter male has browner head markings. Female has greyish-brown head, pale supercilium and ear-coverts, brown eye-stripe to rear of eye and a brown malar stripe. Both sexes have a reddish-brown rump with scaly feather tips.
Voice Song and call is similar to Yellowhammer.
Similar species First winter Yellowhammer has yellowish underparts and yellowish edges to primaries.

Range Breeds in Siberia from the Urals east to the Okhotsk coast, Amur Basin and Sakhalin, south to northern Mongolia and Transbaikalia, also in central China. Winters in Afghanistan, Pakistan, north-western India and northern China.
Status in Japan Rare passage migrant and winter visitor to forest edges and cultivated fields in mountains. Regularly observed on Hegura-jima in autumn.

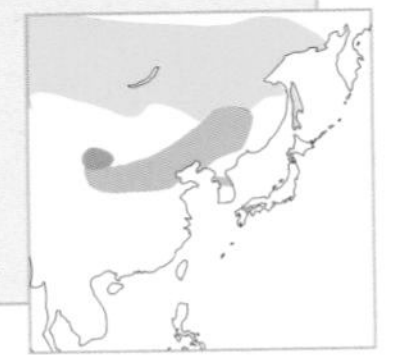

Tadao Shimba

Adult ♂ non-breeding. October, Hegura-jima

Tadao Shimba

Adult ♀ non-breeding. October, Hegura-jima

Tadao Shimba

Juvenile. October, Hegura-jima

Meadow Bunting

Emberiza cioides 16.5cm

Kazuyasu Kisaichi

Adult ♂ *ciopsis*. January, Gunma

Description Brown bunting with distinctive black and white facial pattern. Male *ciopsis* has reddish-brown crown and back with dark streaks, prominent supercilium, black ear-coverts and white moustachial stripe. White throat and pale-brown underparts. Non-breeding male less rufous. Female is duller, with pale-brown supercilium and dark-brown ear-coverts. Male *castaneiceps* has chestnut ear-coverts and mantle.
Voice Song is a clear and accelerating *tschiu chiuhu chichi tsetssere* often delivered from the top of a tree or telephone wires, call is a rapid, thin *zit-zit-zit*.

Kazuyasu Kisaichi

Adult ♀ *ciopsis*. January, Gunma

Range Breeds in eastern Asia from Altai Mountains and Transbaikalia east to Ussuriland, south through Korea to eastern China. Mostly sedentary, but some northern birds move to warmer areas.
Status in Japan The most common bunting. Breeds in forest edges and cultivated fields with thickets. Race *castaneiceps* is a rare visitor from Korea.

Hyuntae Kim

Adult ♂ *castaneiceps*. May, Sosan, South Korea

Japanese Reed Bunting

Emberiza yessoensis 14.5cm

Description Breeding male has black hood, reddish-brown back with dark streaks, greyish lesser-coverts, reddish-brown rump and dark-brown tail with white outer tail feathers. Female has dark brown head, pale supercilium, dark-brown ear-coverts and malar stripe. Non-breeding male is similar to female but has a darker head.
Voice Song is similar to Meadow Bunting and delivered from reeds. Call is a short and trilled *tit*.
Similar species Female Common Reed Bunting is larger, has brown ear-coverts and greyish rump. Pallas's Reed Bunting is paler overall and has a greyish rump.

Range Breeds locally in Ussuriland, north-eastern China and Japan. Winters mainly in eastern China and southern Japan.
Status in Japan Uncommon resident. Breeds in highland meadows of Mount Aso in Kyushu and in reedbeds from central to northern Honshu. Winters in warmer areas.

Tokio Sugiyama

Adult ♂ breeding. June, Ibaragi

Tokio Sugiyama

Adult ♂ non-breeding. March, Aichi

Tokio Sugiyama

Adult ♂ non-breeding. March, Aichi

Tristram's Bunting

Emberiza tristrami 15cm

Akira Yamamoto

Adult ♂ breeding. May, Hegura-jima

Tadao Shimba

First-summer ♀. May, Hegura-jima

Description Brown bunting with distinctive black and white striped head. Male has black head and throat with white crown-stripe, white supercilium and white malar stripe with white spot on ear-coverts. Also a greyish-brown back with dark streaks, reddish-brown rump and tail and brown breast and flanks with dark-brown streaks; lower belly white. Female has dark-brown head with buff-coloured crown-stripe and supercilium. Both sexes have white outer-tail feathers.
Voice Call is a short and metallic *tsit*.
Similar species Rustic Bunting has short crest, whitish underparts and reddish-brown back with scaly, buff-coloured feather edges.

Range Breeds in the Amur Basin, Ussuriland and northern China; migrates south through north-eastern China and Korea. Winters in south-eastern China.
Status in Japan Rare passage migrant to coastal areas along the Sea of Japan. Regularly observed on Tsushima and Hegura-jima in May. Prefers dense cover in wooded areas.

Tokio Sugiyama

Adult ♂ breeding. May, Hegura-jima

Chestnut-eared Bunting

Emberiza fucata 16cm

Description Breeding male has a grey head with conspicuous chestnut ear-coverts, reddish-brown back with dark streaks, white throat and breast with dark malar stripe and breast-band, whitish underparts with dark-streaked flanks. Female and non-breeding male are browner. Both sexes have white outer-tail feathers.
Voice Song is delivered from the top of thickets and is similar to Meadow Bunting in quality, but shorter and rattling. Call is sharp and short *tzit*.
Similar species Little Bunting is smaller, has shorter tail and more striped head.

Range Breeds from north-eastern Mongolia east to Amur Basin and Ussuriland, south to Korea and Japan; also in Himalayas and western China. Winters south to southern China and northern south-east Asia.
Status in Japan Uncommon resident. Breeds in highland grassy meadows and riverbanks from Kyushu northwards. Moves to warmer areas in winter.

Tadao Shimba

Adult ♂ breeding. June, Nagano

Kazuyasu Kisaichi

Adult non-breeding. October, Tobishima

Tadao Shimba

Adult ♂ breeding. June, Nagano

Little Bunting

Emberiza pusilla 12.5cm

Tadao Shimba

Adult breeding. May, Hegura-jima

Kazuyasu Kisaichi

Adult breeding. May, Tsushima

Description Small bunting with small bill and white eye-ring. Chestnut crown-stripe bordered by black stripes, chestnut supercilium and black-bordered chestnut ear-coverts. Greyish-brown back with dark streaks, two indistinct wing-bars and whitish underparts with heavy dark streaks. White outer tail feathers. In non-breeding plumage has cream-coloured crown-stripe and supercilium. Sexes similar.
Voice Call is a thin and sharp *tsit*.
Similar species Chestnut-eared Bunting is larger with greyish head and lacks crown-stripe and supercilium.

Range Breeds in northern Eurasia from northern Scandinavia east through northern Russia to Okhotsk coast. Winters in southern China and northern south-east Asia.
Status in Japan Rare passage and winter visitor to cultivated and grassy fields. Uncommon spring migrant to offshore islands in Sea of Japan.

Kazuyasu Kisaichi

Adult non-breeding. October, Tobishima

Yellow-browed Bunting

Emberiza chrysophrys 15.5cm

Description Male has black head, white crown-stripe, white spot on ear-coverts and a prominent yellow supercilium, whiter to rear of eye. Also a greyish-brown back with dark streaks, reddish-brown rump and whitish underparts with fine dark streaks on breast and flanks. Female is browner.
Voice Call is a sharp and metallic *tsit.*
Similar species Yellow-throated Bunting has short crest and grey rump.

Kazuyasu Kisaichi

Adult ♂. April, Tsushima

Range Breeds in central regions of eastern Siberia south to Lake Baikal area and winters in south-eastern China. Uncommon passage migrant to north-eastern China and Korea.
Status in Japan Rare passage migrant mainly to offshore islands in the Sea of Japan. Frequents forest edges and cultivated fields with thickets.

Kazuyasu Kisaichi

Adult ♀. May, Hegura-jima

Tokio Sugiyama

Adult ♀. May, Tsushima

Rustic Bunting

Emberiza rustica 15cm

Tadao Shimba

Adult ♂ breeding. May, Hegura-jima

Tadao Shimba

Adult ♂ non-breeding. March, Aichi

Description Breeding male has black head with distinctive broad white supercilium, short crest and chestnut nape. Brown above with dark streaks, reddish-brown rump with scaly buff-coloured feather tips and whitish below with broad rusty streaks on breast and flanks. Female and non-breeding male are browner.

Voice Song is usually delivered from the canopy of a tree; a longish, metallic and melodious warble. Call is a soft and metallic *tsit*.

Similar species Female Tristram's Bunting has white crown-stripe and a dark brown stripe over buff-coloured supercilium.

Range Breeds in northern Eurasia from Scandinavia east to Kamchatka, south to Lake Baikal and northern Sakhalin. Winters in China, Korea and Japan.

Status in Japan Abundant winter visitor to open forests, cultivated fields and riverbanks.

Tadao Shimba

Adult ♂ non-breeding. March, Aichi

Yellow-throated Bunting

Emberiza elegans 15.5cm

Description Breeding male is brown with distinctive black and yellow head pattern. Short black crest and grey rump, yellow throat, and white underparts with black breast-patch and brown streaks on flanks. Winter male is duller. Female is duller with pale brown crest and pale yellowish supercilium and breast.
Voice Call is a sharp and metallic *tsit tsit.*
Similar species Non-breeding Rustic Bunting has whiter supercilium, darker malar stripe, reddish-brown rump and shorter tail.

Range Breeds in north-eastern Asia from the Amur Basin and Ussuriland, south to Korea; also in central China. Winters south to southern China and northern south-east Asia.
Status in Japan Uncommon winter visitor to open forests and forest edges. A few breed in mountain forests on Tsushima and in Hiroshima.

Akira Yamamoto

Adult ♂ breeding. January, Aichi

Tadao Shimba

♀. February, Aichi

Tadao Shimba

♂ non-breeding. October, Hegura-jima

Yellow-breasted Bunting

Emberiza aureola 14cm

Mitsuo Imai

Adult ♂ breeding. May, Yonaguni-jima

Description Breeding male has black face and throat, chestnut-brown crown and back and dark-brown wings with a large white patch on wing-coverts. Bright yellow breast and dark-brown tail. Female has brown back with dark streaks, pale supercilium, dark-bordered, brown ear-coverts and yellowish underparts. Both sexes have white outer tail feathers.
Voice Song is a loud and melodious warble. Call is a short, metallic *tsit-tsit*.
Similar species Female Chestnut Bunting is slightly smaller with paler brownish upperparts, indistinct facial pattern and lacks white outer-tail feathers.

Kazuyasu Kisaichi

Adult ♂ breeding. June, Hokkaido

Range Breeds in northern Eurasia from Finland east to far north-eastern Russia and Kamchatka, south to north-eastern China. Winters south to southern China and northern south-east Asia.
Status in Japan Summer visitor to Hokkaido, breeding in grassy meadows and freshwater marshes with shrubs. Common summer visitor until 1980s but population has seriously declined since and only few pairs breed in northern Hokkiado.

Manabu Totsuka

Adult ♀ breeding. June, Hokkaido

Chestnut Bunting

Emberiza rutila 13.5cm

Description Breeding male has rufous head, throat, breast and back; dark-brown wings and tail and a yellow belly with brown streaks on flanks. Non-breeding male has pale feather edging above. Female has brown back with dark-brown streaks, pale yellowish head with streaked crown, indistinct eye-stripe and malar stripe, pale yellowish underparts with dark streaks. Both sexes lack white outer-tail feathers.
Voice Call is a short and metallic *tit*.
Similar species Female Yellow-breasted Bunting has pale supercilium, dark-bordered pale ear-coverts, yellow underparts and white outer-tail feathers.

Range Breeds in southern Siberia from Lake Baikal east to the Okhotsk coast and lower Amur Basin, south to northern Ussuriland and northern China. Winters in southern China and northern south-east Asia. Uncommon passage migrant to Korea.
Status in Japan Rare passage migrant, mainly in spring, to offshore islands in Sea of Japan. Cultivated fields and grassy fields with thickets.

Tadao Shimba

First-summer ♂. May, Hegura-jima

Tadao Shimba

Adult ♀. May, Hegura-jima

Yoshio Goto

Adult ♂ non-breeding. October, Hegura-jima

Japanese Yellow Bunting

Emberiza sulphurata 14cm

Norio Yamagata

First-summer ♂. May, Hegura-jima

Tadao Shimba

Adult ♀. May, Hegura-jima

Description Male has greenish-grey head with dark lores and white eye-ring. Greenish-brown back with dark streaks and two prominent white wing-bars. Yellow throat and underparts with brown streaks on flanks. Female is duller with pale yellowish underparts. Both sexes have white outer-tail feathers.

Voice Song is similar to Black-faced Bunting but longer and flutier. Call is a soft and thin *tsit*.

Similar species Male Black-faced Bunting has darker head and lacks white eye-ring. Female Black-faced Bunting has pale supercilium, dark malar stripe and heavily streaked underparts.

Range Breeds in Japan and winters in southern China, Taiwan and the Philippines.

Status in Japan Locally uncommon summer visitor to open deciduous forests and forest edges in central and northern Honshu. Rare passage migrant elsewhere. A few winter in southern Japan.

Tadao Shimba

Adult ♂ breeding. May, Hegura-jima

Black-faced Bunting

Emberiza spodocephala 16cm

Description Two races in Japan; male *personata* has dark greenish-grey head with dark lores and a brown back with dark streaks. Yellowish throat and underparts with dark-brown streaks on breast and flanks. Female has brown head with pale yellowish supercilium and dark bordered, greenish-brown ear-coverts. Pale-brown underparts with heavily streaked breast and flanks. Male becomes duller in winter. Both sexes have white outer-tail feathers. Race *spodocephala* has dark-greyish hood and breast, and pale yellow underparts.
Voice Song is delivered from exposed perch; slow and melodious *chip-chu chi-chi-chiriririr.* Call is a strong and harsh *zit*.
Similar species Japanese Yellow Bunting is smaller, and has paler face with white eye-ring and less streaked underparts.

Range Breeds in Siberia from the Altai east to the Okhotsk coast, lower Amur Basin and Sakhalin, south to north-eastern China, northern Korea and Japan; also in central China. Winters south to southern China, Taiwan and southern Japan.
Status in Japan Breeds in deciduous and mixed forests from central Honshu northwards. Moves to warmer areas in winter and is common in city parks and gardens with undergrowth. Race *spodocephala* breeds in Siberia and is a rare passage migrant mainly to Kyushu and the Ryukyu Islands.

Tokio Sugiyama

Adult ♂ *personata*. March, Aichi

Tadao Shimba

First-winter ♂ *personata*. February, Aichi

Yoshio Goto

Adult ♂ *spodocephala*. May, Hegura-jima

Grey Bunting

Emberiza variabilis 16.5cm

Tadao Shimba

First-summer ♂. May, Hegura-jima

Tadao Shimba

Adult ♀. February, Aichi

Description Breeding male is dark slate-grey overall with darker streaks on back and whitish underparts. Winter male has a browner back. Female has brown face, pale supercilium and dark malar stripe, white throat, brown back with dark streaks and pale underparts. Both sexes have large dark bills with pinkish lower mandible, pinkish legs and lack white on tail feathers.
Voice Song is a loud and melodious three syllable *chewee-tseeh-tseeh.* Call is a soft, thin and metallic *tit.*
Similar species Black-faced Bunting is smaller and has yellowish underparts and white outer-tail feathers.

Range Breeds in southern Kamchatka, Kuril Islands, Sakhalin and northern Japan. Winters in Japan.
Status in Japan Breeds in mountain forests from central Honshu northwards. Locally common winter visitor to warmer areas. Skulks on the forest floor in dense evergreen forests in winter.

Tadao Shimba

Probable second-winter ♂. February, Aichi

Pallas's Reed Bunting

Emberiza pallasi 14cm

Description Similar to Common Reed Bunting, but smaller. Breeding male has buff nape, two pale wing-bars and pale grey lesser coverts. Female and winter male have pale-brown back with dark streaks and bluish-grey lesser-coverts. Pale-greyish rump and uppertail-coverts and brown tail with white outer-tail feathers, dark bill with pinkish lower mandible.

Voice Song is a short series of soft trills. Call is a soft *tsilit* or *tsit*.

Similar species Common Reed Bunting is larger and browner and has grey bill, streaked underparts and a darker rump. Japanese Reed Bunting is browner and has reddish-brown rump.

Range Breeds on tundra in northern Siberia and in mountains in southern Siberia. Winters in eastern China and Korea.

Status in Japan Rare winter visitor mainly to western Japan. Reedbeds and grassy fields.

Akira Yamamoto

Adult ♂ non-breeding. January, Aichi

Akira Yamamoto

Adult ♀ non-breeding. January, Aichi

Tadao Shimba

Adult ♂ non-breeding. February, Aichi

Common Reed Bunting

Emberiza schoeniclus 16cm

Akira Yamamoto

Adult ♂ breeding. June, Hokkaido

Description Breeding male has black head, throat and upper breast contrasting with white of malar stripe, collar and underparts. Reddish-brown above with dark streaks, greyish rump and dark-brown tail. Female has brown crown and ear-coverts, pale supercilum, dark malar stripe and white underparts.
Voice Song is a short series of clear and metallic trilling, *zritt zreet zrrrrh.* Call is a harsh, rising *tseeep*.
Similar species Pallas's Reed Bunting is smaller and paler, and has greyish lesser-coverts and pink lower mandible.

Tadao Shimba

Adult ♀ non-breeding. February, Aichi

Range Breeds in Eurasia from Europe east to the Lena River and Amur Basin, also in Kamchatka, Sakhalin, Kuril Islands and northern Japan. Moves south in winter and a rare winter visitor to Korea.
Status in Japan. Breeds in freshwater marshes in northern Honshu and Hokkaido and moves to warmer areas in winter.

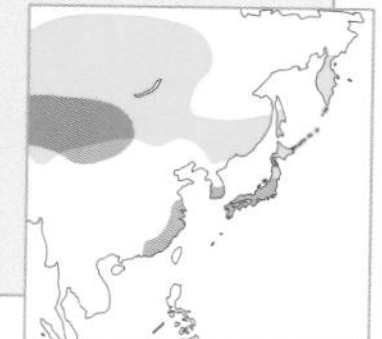

Tadao Shimba

Adult ♂ non-breeding. February, Aichi

Ortolan Bunting

Emberiza hortulana 16.5cm

Description Breeding male has a uniform greenish-grey head with yellowish eye-ring, yellow throat and moustachial-stripe and rufous belly. Greenish-grey back with dark streaks, dark-brown wings with reddish-brown fringes to wing-coverts and flight feathers. Female and winter male have greyish-brown head with fine dark-brown streaks on head, nape and breast. First winter male is similar to female.

Range Breeds in western Eurasia from Europe east to the Altai mountains. Winters in Africa.
Status in Japan Vagrant. Recorded once on Hegura-jima in October 1986.

Tomi Muukkonen

Adult ♂. May, Joutsa, Finland

Black-headed Bunting

Emberiza melanocephala 16.5cm

Description Large bunting with large bill. Male has black face and crown, rusty-brown back and rump, brown tail, dark-brown wings with buff feather edges, and bright yellow throat, sides of neck and underparts. Female has pale greyish-brown back with brown streaks, pale buff-coloured underparts and pale yellowish undertail-coverts.
Voice Call is a soft *tsit-tsit*.

Range Breeds from eastern Mediterranean to Iran and winters in the Indian subcontinent.
Status in Japan Rare passage migrant to cultivated and grassy fields. Observed almost annually despite the distance from its normal range; some may be escapes.

Teruaki Ishi

Adult ♂ breeding. May, Kyoto

White-crowned Sparrow

Zonotrichia leucophrys 16–18cm

Tadao Shimba

Adult. April, Vancouver, Canada

Description Distinctive black and white striped head, bluish-grey face and underparts, greyish back with dark-brown streaks and dark-brown wings. Pale-brown rump and dark-brown notched tail. First-winter is browner with reddish-brown stripes on sides of head.
Similar species First-winter Golden-crowned Sparrow is browner with fine dark-brown streaks and yellowish-brown crown patch.

Range Breeds across northern North America and much of western North America. Winters in southern North America.
Status in Japan Vagrant. Occurs in wooded areas and forest edges.

Golden-crowned Sparrow

Zonotrichia atricapilla 18cm

Akira Yamamoto

Adult. March, Vancouver, Canada

Description Black head with yellow crown, grey face and throat and greyish-brown underparts. Brown back with dark streaks, brown rump and dark-brown tail. First-winter has brown head with fine dark-brown streaks and yellowish-brown crown patch.
Similar species First-winter White-crowned Sparrow has reddish-brown stripe on sides of head and greyish-brown breast.

Range Breeds in north-western North America and winters on west coasts of North America.
Status in Japan Vagrant. Less frequently observed than White-crowned Sparrow.

Savannah Sparrow

Passerculus sandwichensis 14cm

Description Greyish-brown back with dark streaks, dark-brown crown with pale crown-stripe, pale yellowish supercilium and brown ear-coverts bordered with dark brown. White throat with dark malar stripe, whitish underparts with dark-brown streaks on breast and flanks and notched tail with no white on outer feathers.
Voice Call is a thin and metallic *seep*.

Shigeo Ozawa

Adult. Tokyo, November

Range Widespread throughout northern and central North America and winters south to northern South America.
Status in Japan Very rare winter visitor to grassy fields and cultivated fields.

Fox Sparrow

Passerella iliaca 18cm

Description Dark greyish-brown head and back, reddish-brown rump and tail, whitish below with heavy arrowhead-shaped spots on throat, breast and flanks.

Range Breeds across northern North America and much of western North America; also on the Aleutian Islands. Winters in southern North America.
Status in Japan Vagrant. Two records of Aleutian race *unalaschcensis* in Tochigi (1935) and Hokkaido (1985).

Akira Yamamoto

Adult. March, Vancouver, Canada

Lapland Bunting

Calcarius lapponicus 16cm

Tokio Sugiyama

Adult ♂ non-breeding. February, Aichi

Description Breeding male has black head and breast with distinctive white stripe extending back from eye down sides of breast. Also a chestnut nape, brown mantle with dark streaks and white underparts with dark streaks on flanks. Female has buffish supercilium, fine dark streaks on crown, dark borders to buff-coloured ear-coverts and white throat with dark malar stripe. Non-breeding male is similar to female, but has dark crown and dark patch on upper breast.

Voice Call is short, trilled *prrt*.

Teruaki Ishii

Probable juvenile. October, Hegura-jima

Range Breeds on tundra in arctic regions of northern Eurasia, northern North America and Greenland. Winters in coastal areas and on prairies in temperate regions.

Status in Japan Rare winter visitor to Hokkaido and coastal areas around the Sea of Japan. Coastal grassy fields and cultivated fields.

Jari Peltomäki

Adult ♂ breeding, June , Varanger, Norway

Snow Bunting

Calcarius nivalis 16–17cm

Description Breeding male has white head and underparts contrasting with black back. White rump, white tail with dark central feathers and dark wing tips. Female has mottled black and brown crown and back and rusty-brown ear-coverts and breast. Non-breeding male has rusty-brown crown, ear-coverts and breast, mottled brown and black back. In flight, male shows contrasting white wings with black primaries, female has brown wing-coverts and mainly white secondaries.
Voice Call is short and soft *pyu*.

Range Circumpolar. Breeds on islands and coasts in Arctic regions. Winters in coastal areas and prairies of temperate regions.
Status in Japan Uncommon winter visitor to northern Japan. Frequents coastal areas and grassy fields.

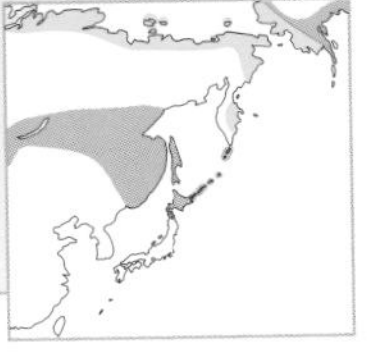

Kazuyasu Kisaichi

First-winter ♂. November, Hegura-jima

Augusto Faustino

Adult ♂ breeding. July, Chukotka, Russia

Tokio Sugiyama

Adult ♂ non-breeding. February, Ishikawa

Tadao Shimba

Narcissus Flycatcher. May, Hegura-jima

APPENDICES

APPENDICES A–C SHOW SPECIES that have been recorded in the region; Appendix A shows rarer vagrants that are not included in the main section of the book; Appendix B shows species that have become extinct from the region, or are presumed to be so; and Appendix C shows species that have been extirpated from the region but that survive elsewhere.

J = recorded in Japan
K = recorded in Korea
C = recorded in north-east China
R = recorded in north-east Russia

Appendix A

Rarer vagrant species that do not appear in the main text

English Name	Scientific Name	J	K	C	R
Great Northern Diver	*Gavia immer*	✓			✓
Dark-rumped Petrel	*Pterodroma phaeopygia*	✓			
Stejneger's Petrel	*Pterodroma longirostris*	✓			
Juan Fernandez Petrel	*Pterodroma externa*	✓			
Christmas Island Shearwater	*Puffinus nativitatis*	✓			
Little Shearwater	*Puffinus assimilis*	✓			
Manx Shearwater	*Puffinus puffinus*	✓			
Newell's Shearwater	*Puffinus newelll*	✓			
Lesser Whistling Duck	*Dendrocygna javanica*	✓			
Cotton Pygmy Goose	*Nettapus cormomandelianus*	✓			
Philippine Duck	*Anas luzonica*	✓			
Hooded Merganser	*Mergus cucullatus*	✓			
Great Blue Heron	*Ardea herodias*				✓
Black Bittern	*Ixobrychus flavicollis*	✓	✓		
Saker Falcon	*Falco cherrug*		✓		

English Name	Scientific Name	J	K	C	R
Black Baza	*Aviceda leuphotes*				✓
Black-winged Kite	*Elanus caeruleus*	✓			
Bald Eagle	*Haliaeetus leucocephalus*				✓
Pallas's Fish Eagle	*Haliaeetus leucoryphrus*			✓	
Bearded Vulture	*Gypaetus barbatus*		✓		
Short-toed Eagle	*Circaetus gallicus*				✓
Western Marsh Harrier	*Circus aeruginosus*	✓	✓		
Steppe Eagle	*Aquila rapax*		✓		✓
Little Bustard	*Tetrax tetrax*	✓			
American Black Oystercatcher	*Haematopus bachmani*				✓
Semipalmated Plover	*Charadrius semipalmatus*	✓			✓
Slender-billed Curlew	*Numenius tenuirostris*	✓			
Short-billed Dowitcher	*Limnodromus griseus*	✓			
Hudsonian Godwit	*Limosa haemastica*	✓			
Killdeer	*Charadrius vociferus*				✓
Jack Snipe	*Lymnocryptes minimus*	✓	✓	✓	✓
Solitary Sandpiper	*Tringa solitaria*				✓
White-rumped Sandpiper	*Calidris fuscicollis*	✓			
Black Turnstone	*Arenaria melanocephala*				✓
Iceland Gull	*Larus glaucoides*	✓	✓		
Brown-headed Gull	*Larus brunnicephalus*	✓			
Lesser Crested Tern	*Sterna bengalensis*	✓			
Grey Noddy	*Procelsterna cerulea*	✓			
Black Guillemot	*Cepphus grylle*				✓
Cassin's Auklet	*Ptychoramphus aleuticus*				✓
Chukar	*Alectoris chukar*			✓	
Stock Dove	*Columba oenas*	✓	✓		
Chestnut-winged Cuckoo	*Clamator coromandus*	✓	✓		

English Name	Scientific Name	J	K	C	R
Large Hawk Cuckoo	*Cuculus sparverioides*	✓			
Asian Drongo Cuckoo	*Surniculus lugubris*	✓			
Black-and-white Cuckoo	*Cuculus jacobinus*	✓			
Asian Koel	*Eudnamys scolopaceus*	✓			
Lesser Coucal	*Centropus bengalensis*	✓			
Grass Owl	*Tyto capensis*	✓			
Himalayan Swiftlet	*Collocallia brevirostris*	✓			
Rufous Hummingbird	*Selasphorus rufus*				✓
White-throated Kingfisher	*Halcyon smyrnensis*	✓			
Collared Kingfisher	*Todiramphus chloris*	✓			
Rainbow Bee-eater	*Merops ornatus*	✓			
White-bellied Woodpecker	*Dryocopos javensis*	✓	✓		
Hooded Pitta	*Pitta sordida*	✓			
Black-winged Cuckoo-shrike	*Coracina melaschistos*	✓	✓		
Plain Martin	*Riparia paludicola*	✓			
Northern House Martin	*Delichon urbica*	✓	✓		
Tree Swallow	*Tachycineta bicolor*	✓			✓
Cliff Swallow	*Petrochelidon albifrons*				✓
Mongolian Lark	*Melanocorypha mongolica*	✓		✓	
Tickell's Leaf Warbler	*Phylloscopus affinis*	✓			
Hume's Leaf Warbler	*Phylloscopus humei*		✓		
Crested Myna	*Acridotheres cristatellus*	✓			
Plumbeous Water Redstart	*Rhyacornis fuliginosus*		✓		
White-tailed Robin	*Cinclidium leucurum*	✓			
Whinchat	*Saxicola rubetra*	✓			
Common Rock Thrush	*Monticola saxatilis*	✓			
Spotted Flycatcher	*Muscicapa striata*	✓			
Ferruginous Flycatcher	*Muscicapa ferruginea*	✓			

English Name	Scientific Name	J	K	C	R
Vivid Niltava	*Niltava vivida*	✓			
Rufous-billed Niltava	*Niltava sundara*	✓			
Rosy Pipit	*Anthus roseatus*		✓		
Water Pipit	*Anthus spinoletta*		✓		
Red-backed Shrike	*Lanius collurio*	✓			
Ashy Drongo	*Dicurus leucophaeus*	✓			
Chaffinch	*Fringilla coelebs*	✓			
Grey-crowned Rosy Finch	*Leucosticte tephrocotis*				✓
Wilson's Warbler	*Wilsonia pusilla*	✓			
Orange-crowned Warbler	*Vermivora celata*				✓
Yellow-rumped Warbler	*Dendroica coronata*				✓
Northern Waterthrush	*Seirus noveboracensis*				✓
Rusty Blackbird	*Euphagus carolinensis*				✓
Western Meadowlark	*Sturnella neglecta*				✓
Crested Bunting	*Melophus lathami*	✓			
Grey-necked Bunting	*Emberiza buchanani*	✓			
Red-headed Bunting	*Emberiza bruniceps*	✓	✓		
Dark-eyed Junco	*Junco hyemalis*				✓
American Tree Sparrow	*Spizella arborea*				✓

Appendix B

Species extinct, or presumed extinct

English Name	Scientific Name	J	K	C	R
Spectacled Cormorant	*Phalacrocorax perspicillatus*				✓
Crested Shelduck	*Tadorna cristata*	✓	✓		✓
Bonin Woodpigeon	*Columba versicolor*	✓			
Ryukyu Woodpigeon	*Columba jouyi*	✓			
Miyako Island Kingfisher	*Todiramphus miyakoensis*	✓			
Bonin Islands Thrush	*Cichlopasser terrestris*	✓			
Bonin Islands Grosbeak	*Chaunoproctus ferreorostris*	✓			

Appendix C

Species extirpated from the region

English Name	Scientific Name	J	K	C	R
Rufous Night Heron	*Nycticorax caledonicus*	✓			
Crested Ibis	*Nipponia nippon*	✓	✓	✓	✓
White-browed Crake	*Porzana cinereus*	✓			

Appendix D

THE TABLE BELOW IS INCLUDED as a useful tool for birdwatchers visiting Japan, allowing them to communicate what they are seeing to local birders and perhaps non-birders, too. All Japanese people will know words such as *suzume* (sparrow), *tsubame* (swallow) and *karasu* (crow), and even more specific names like *kawasemi* (Common Kingfisher) or *karugamo* (Spot-billed Duck) will be familiar to most.

English name	Japanese name
Red-throated Diver	Abi
Black-throated Diver	Ohamu
Pacific Diver	Shiroeri-ohamu
White-billed Diver	Hashijiro-abi
Little Grebe	Kaitsuburi
Red-necked Grebe	Akaeri-kaitsuburi
Great Crested Grebe	Kanmuri-kaitsuburi
Black-necked Grebe	Hajiro-kaitsuburi
Slavonian Grebe	Mimi-kaitsuburi
Short-tailed Albatross	Ahoudori
Laysan Albatross	Ko-ahoudori
Black-footed Albatross	Kuroashi-ahoudori
Northern Fulmar	Furuma-kamome
Providence Petrel	Hajiro-mizunagidori
White-necked Petrel	O-shirohara-mizunagidori
Bonin Petrel	Shirohara-mizunagidori
Streaked Shearwater	O-mizunagidori
Flesh-footed Shearwater	Akaashi-mizunagidori
Wedge-tailed Shearwater	Onaga-mizunagidori
Buller's Shearwater	Minami-onaga-mizunagidori
Sooty Shearwater	Haiiro-mizunagidori
Short-tailed Shearwater	Hashiboso-mizunagidori
Bulwer's Petrel	Anadori
Audubon's Shearwater	Seguro-mizunagidori

English name	Japanese name
Maderian Storm-petrel	Kuro-koshijiro-umitsubame
Leach's Storm-petrel	Koshijiro-umitsubame
Swinhoe's Storm-petrel	Hime-umitsubame
Tristram's Storm-petrel	Ohsuton-umitsubame
Matsudaira's Storm-petrel	Kuro-umitsubame
Fork-tailed Storm-petrel	Haiiro-umitsubame
Brown Booby	Katsuo-dori
Dalmatian Pelican	Haiiro-perikan
Great Cormorant	Kawa-u
Japanese Cormorant	Umi-u
Pelagic Cormorant	Hime-u
Red-faced Cormorant	Chishima-ugarasu
Grey Heron	Ao-sagi
Purple Heron	Murasaki-sagi
Great Egret	Dai-sagi
Intermediate Egret	Chu-sagi
Chinese Egret	Karashira-sagi
Little Egret	Ko-sagi
Pacific Reef Egret	Kuro-sagi
Chinese Pond Heron	Akagashira-sagi
Cattle Egret	Ama-sagi
Striated Heron	Sasagoi
Black-crowned Night Heron	Goi-sagi
Japanese Night Heron	Mizogoi

English name	Japanese name
Malayan Night Heron	Zuguro-mizogoi
Yellow Bittern	Yoshigoi
Schrenck's Bittern	O-yoshigoi
Cinnamon Bittern	Ryukyu-yoshigoi
Eurasian Bittern	Sankanogoi
Black Stork	Nabekou
Oriental Stork	Kounotori
Black-headed Ibis	Kurotoki
Eurasian Spoonbill	Herasagi
Black-faced Spoonbill	Kurotsura-herasagi
Mute Swan	Kobu-hakucho
Trumpeter Swan	Naki-hakucho
Whooper Swan	O-hakucho
Tundra Swan	Ko-hakucho
Swan Goose	Sakatsura-gan
Bean Goose	Hishikui
Greater White-fronted Goose	Ma-gan
Lesser White-fronted Goose	Karigane
Snow Goose	Haku-gan
Greylag Goose	Haiiro-gan
Cackling Goose	Shijukara-gan
Brent Goose	Koku-gan
Ruddy Shelduck	Akatsukushi-gamo
Common Shelduck	Tsukushi-gamo
Mandarin Duck	Oshidori
Eurasian Wigeon	Hidori-gamo
American Wigeon	Amerika-hidori
Mallard	Ma-gamo
Falcated Duck	Yoshi-gamo
Gadwall	Okayoshi-gamo
Baikal Teal	Tomoe-gamo
Common Teal	Ko-gamo
Spot-billed Duck	Karu-gamo
Northern Pintail	Onaga-gamo
Garganey	Shimaaji
Northern Shoveler	Hashibiro-gamo
Red-crested Pochard	Akahashi-hajiro
Common Pochard	Hoshi-hajiro
Canvasback	O-hoshi-hajiro

English name	Japanese name
Ring-necked Duck	Kubiwa-kinkuro
Ferruginous Duck	Mejiro-gamo
Baer's Pochard	Aka-hajiro
Tufted Duck	Kinkuro-hajiro
Greater Scaup	Suzu-gamo
Lesser Scaup	Ko-suzu-gamo
Steller's Eider	Kokewata-gamo
Harlequin Duck	Shinori-gamo
Long-tailed Duck	Kohri-gamo
Black Scoter	Kuro-gamo
White-winged Scoter	Birohdo-kinkuro
Common Goldeneye	Hohjiro-gamo
Bufflehead	Hime-hajiro
Smew	Miko-aisa
Red-breasted Merganser	Umi-aisa
Goosander	Kawa-aisa
Scaly-sided Merganser	Kourai-aisa
Osprey	Misago
Oriental Honey-buzzard	Hachikuma
Black Kite	Tobi
White-tailed Eagle	Ojiro-washi
Steller's Sea Eagle	O-washi
Eurasian Black Vulture	Kuro-hagewashi
Crested Serpent Eagle	Kanmuri-washi
Eastern Marsh Harrier	Chuhi
Hen Harrier	Haiiro-chuhi
Pied Harrier	Madara-chuhi
Chinese Goshawk	Akahara-daka
Japanese Sparrowhawk	Tsumi
Eurasian Sparrowhawk	Hai-taka
Northern Goshawk	O-taka
Grey-faced Buzzard	Sashiba
Common Buzzard	Nosuri
Upland Buzzard	O-nosuri

English name	Japanese name
Rough-legged Buzzard	Keashi-nosuri
Greater Spotted Eagle	Karafuto-washi
Eastern Imperial Eagle	Katashiro-washi
Golden Eagle	Inu-washi
Mountain Hawk Eagle	Kuma-taka
Eurasian Kestrel	Chougenbo
Amur Falcon	Akaashi-chougenbo
Merlin	Ko-chougenbo
Northern Hobby	Chigo-hayabusa
Gyrfalcon	Shiro-hayabusa
Peregrine Falcon	Hayabusa
Rock Ptarmigan	Raicho
Hazel Grouse	Ezo-raicho
Japanese Quail	Uzura
Chinese Bamboo Partridge	Kojukei
Common Pheasant	Kourai-kiji
Green Pheasant	Kiji
Copper Pheasant	Yamadori
Barred Buttonquail	Mifu-uzura
Demoiselle Crane	Aneha-zuru
Common Crane	Kuro-zuru
Siberian Crane	Sodeguro-zuru
Sandhill Crane	Kanada-zuru
White-naped Crane	Mana-zuru
Hooded Crane	Nabe-zuru
Red-crowned Crane	Tancho
Slaty-legged Crake	O-kuina
Swinhoe's Rail	Shima-kuina
Water Rail	Kuina
White-breasted Waterhen	Shirohara-kuina
Okinawa Rail	Yanbaru-kuina
Baillon's Crake	Hime-kuina
Ruddy-breasted Crake	Hi-kuina
Watercock	Tsuru-kuina
Common Moorhen	Ban

English name	Japanese name
Common Coot	O-ban
Great Bustard	Nogan
Pheasant-tailed Jacana	Renkaku
Greater Painted-snipe	Tama-shigi
Eurasian Oystercatcher	Miyakodori
Black-winged Stilt	Seitaka-shigi
Oriental Pratincole	Tsubame-chidori
Pied Avocet	Sorihashi-seitaka-shigi
Northern Lapwing	Tageri
Grey-headed Lapwing	Keri
Pacific Golden Plover	Munaguro
Grey Plover	Daizen
Common Ringed Plover	Hajiro-ko-chidori
Long-billed Plover	Ikaru-chidori
Little Ringed Plover	Ko-chidori
Kentish Plover	Shiro-chidori
Lesser Sand Plover	Medai-chidori
Greater Sand Plover	O-medai-chidori
Oriental Plover	O-chidori
Eurasian Dotterel	Kobashi-chidori
Eurasian Woodcock	Yama-shigi
Amami Woodcock	Amami-yama-shigi
Solitary Snipe	Ao-shigi
Latham's Snipe	Oji-shigi
Pintail Snipe	Hario-shigi
Swinhoe's Snipe	Chuji-shigi
Common Snipe	Ta-shigi
Long-billed Dowitcher	Ohashi-shigi
Asian Dowitcher	Shiberia-ohashi-shigi
Black-tailed Godwit	Oguro-shigi
Bar-tailed Godwit	O-sorihashi-shigi
Little Curlew	Koshaku-shigi
Bristle-thighed Curlew	Harimomo-chushaku

English name	Japanese name
Whimbrel	Chushaku-shigi
Eurasian Curlew	Daishaku-shigi
Far-eastern Curlew	Houroku-shigi
Spotted Redshank	Tsuru-shigi
Common Redshank	Akaashi-shigi
Marsh Sandpiper	Ko-aoashi-shigi
Common Greenshank	Aoashi-shigi
Nordmann's Greenshank	Karafuto-aoashi-shigi
Greater Yellowlegs	O-kiashi-shigi
Lesser Yellowlegs	Ko-kiashi-shigi
Green Sandpiper	Kusa-shigi
Wood Sandpiper	Takabu-shigi
Terek Sandpiper	Sorihashi-shigi
Common Sandpiper	Iso-shigi
Grey-tailed Tattler	Kiashi-shigi
Wandering Tattler	Meriken-kiashi-shigi
Ruddy Turnstone	Kyojo-shigi
Great Knot	Oba-shigi
Red Knot	Ko-oba-shigi
Sanderling	Miyubi-shigi
Western Sandpiper	Hime-hama-shigi
Baird's Sandpiper	Hime-uzura-shigi
Red-necked Stint	Tounen
Little Stint	Yohroppa-tounen
Temminck's Stint	Ojiro-tounen
Long-toed Stint	Hibari-shigi
Pectoral Sandpiper	Amerika-uzura-shigi
Sharp-tailed Sandpiper	Uzura-shigi
Curlew Sandpiper	Saruhama-shigi
Dunlin	Hama-shigi
Rock Sandpiper	Chishima-shigi
Spoon-billed Sandpiper	Hera-shigi
Broad-billed Sandpiper	Kiriai
Ruff	Erimaki-shigi
Buff-breasted Sandpiper	Komon-shigi
Grey Phalarope	Haiiro-hireashi-shigi

English name	Japanese name
Red-necked Phalarope	Akaeri-hireashi-shigi
South Polar Skua	O-touzokukamome
Pomarine Skua	Touzokukamome
Arctic Skua	Kuro-touzoku-kamome
Long-tailed Skua	Shirohara-touzokukamome
Black-tailed Gull	Umineko
Common Gull	Kamome
Glaucous-winged Gull	Washi-kamome
Glaucous Gull	Shiro-kamome
Thayer's Gull	Kanada-kamome
Slaty-backed Gull	O-seguro-kamome
Heuglin's Gull	Hoigurin-kamome
Pallas's Gull	O-zuguro-kamome
Vega Gull	Seguro-kamome
Mongolian Gull	Mongoru-kamome
Black-headed Gull	Yuri-kamome
Saunders's Gull	Zuguro-kamome
Ivory Gull	Zouge-kamome
Ross's Gull	Hime-kubiwa-kamome
Black-legged Kittiwake	Mitsuyubi-kamome
Gull-billed Tern	Hashibuto-ajisashi
Roseate Tern	Beni-ajisashi
Great Crested Tern	O-ajisashi
Caspian Tern	Oni-ajisashi
Black-naped Tern	Eriguro-ajisashi
Common Tern	Ajisashi
Arctic Tern	Kyoku-ajisashi
Little Tern	Ko-ajisashi
Aleutian Tern	Koshijiro-ajisashi
Sooty Tern	Seguro-ajisashi
Bridled Tern	Mamijiro-ajisashi
White-winged Tern	Hajiro-kurohara-ajisashi
Whiskered Tern	Kurohara-ajisashi
Black Tern	Hashiguro-kurohara-ajisashi

English name	Japanese name
Black Noddy	Hime-kuro-ajisashi
Brown Noddy	Kuro-ajisashi
Common Guillemot	Umigarasu
Brünnich's Guillemot	Hashibuto-umigarasu
Spectacled Guillemot	Keimafuri
Pigeon Guillemot	Umibato
Long-billed Murrelet	Madara-umisuzume
Japanese Murrelet	Kanmuri-umisuzume
Ancient Murrelet	Umisuzume
Least Auklet	Ko-umisuzume
Parakeet Auklet	Umi-oumu
Crested Auklet	Etorofu-umisuzume
Whiskered Auklet	Ko-umisuzume
Rhinoceros Auklet	Utou
Horned Puffin	Tsunomedori
Tufted Puffin	Etopirika
Japanese Wood Pigeon	Karasu-bato
Oriental Turtle Dove	Kiji-bato
Eurasian Collared Dove	Shirako-bato
Red Turtle Dove	Beni-bato
Emerald Dove	Kin-bato
White-bellied Green Pigeon	Ao-bato
Whistling Green Pigeon	Zuaka-aobato
Indian Cuckoo	Seguro-kakkou
Oriental Cuckoo	Tsutsudori
Little Cuckoo	Hototogisu
Northern Hawk Cuckoo	Ju-ichi
Common Cuckoo	Kakkou
Collared Scops Owl	O-konohazuku
Ryukyu Scops Owl	Ryukyu-konohazuku
Oriental Scops Owl	Konoha-zuku
Snowy Owl	Shiro-fukurou
Tengmalm's Owl	Kinme-fukurou

English name	Japanese name
Blakiston's Fish Owl	Shima-fukurou
Ural Owl	Fukurou
Brown Hawk Owl	Aoba-zuku
Eurasian Eagle Owl	Washi-mimizuku
Short-eared Owl	Komimi-zuku
Long-eared Owl	Torafu-zuku
White-throated Needletail	Hario-amatsubame
Pacific Swift	Ama-tsubame
House Swift	Hime-amatsubame
Common Kingfisher	Kawasemi
Ruddy Kingfisher	Aka-shoubin
Black-capped Kingfisher	Yama-shoubin
Crested Kingfisher	Yamasemi
Grey Nightjar	Yotaka
Dollarbird	Buppousoh
Eurasian Wryneck	Arisui
Japanese Pygmy Woodpecker	Ko-gera
Great Spotted Woodpecker	Aka-gera
Lesser Spotted Woodpecker	Ko-akagera
White-backed Woodpecker	O-akagera
Black Woodpecker	Kuma-gera
Japanese Green Woodpecker	Ao-gera
Grey-headed Woodpecker	Yama-gera
Three-toed Woodpecker	Miyubi-gera
Okinawa Woodpecker	Noguchi-gera
Eurasian Hoopoe	Yatsugashira
Fairy Pitta	Yairocho
Greater Short-toed Lark	Hime-koutenshi
Asian Short-toed Lark	Ko-hibari
Eurasian Skylark	Hibari

English name	Japanese name
Horned Lark	Hama-hibari
Sand Martin	Shoudou-tsubame
Asian House Martin	Iwa-tsubame
Barn Swallow	Tsubame
Pacific Swallow	Ryukyu-tsubame
Red-rumped Swallow	Koshiaka-tsubame
Forest Wagtail	Iwami-sekirei
White Wagtail	Haku-sekirei
Japanese Wagtail	Seguro-sekirei
Yellow Wagtail	Tsumenaga-sekirei
Grey Wagtail	Ki-sekirei
Citrine Wagtail	Kigashira-sekirei
Meadow Pipit	Makiba-tahibari
Richard's Pipit	Mamijiro-tahibari
Blyth's Pipit	Ko-mamijiro-tahibari
Pechora Pipit	Sejiro-tahibari
Olive-backed Pipit	Binzui
Red-throated Pipit	Muneaka-tahibari
Buff-bellied Pipit	Ta-hibari
Tree Pipit	Yohroppa-binzui
Light-vented Bulbul	Shirogashira
Brown-eared Bulbul	Hiyodori
Ashy Minivet	Sanshoukui
Goldcrest	Kiku-itadaki
Bohemian Waxwing	Ki-renjaku
Japanese Waxwing	Hi-renjaku
Brown Dipper	Kawa-garasu
Winter Wren	Misosazai
Alpine Accentor	Iwa-hibari
Siberian Accentor	Yama-hibari
Japanese Accentor	Kayakuguri
Blue Rock Thrush	Iso-hiyodori
Siberian Thrush	Mamijiro
White's Thrush	Tora-tsugumi
Amami Thrush	O-tora-tsugumi
Grey-backed Thrush	Kara-akahara
Japanese Thrush	Kuro-tsugumi
Eyebrowed Thrush	Mamichajinai
Brown-headed Thrush	Akahara

English name	Japanese name
Pale Thrush	Shirohara
Izu Islands Thrush	Akakokko
Dusky Thrush	Tsugumi
Zitting Cisticola	Sekka
Asian Stubtail	Yabusame
Japanese Bush Warbler	Uguisu
Lanceolated Warbler	Makino-sennyu
Middendorff's Grass-hopper Warbler	Shima-sennyu
Styan's Grasshopper Warbler	Uchiyama-sennyu
Gray's Grasshopper Warbler	Ezo-sennyu
Pallas's Grasshopper Warbler	Shiberia-sennyu
Japanese Marsh Warbler	O-sekka
Black-browed Reed Warbler	Ko-yoshikiri
Oriental Reed Warbler	O-yoshikiri
Dusky Warbler	Muji-sekka
Radde's Warbler	Karafuto-muji-sekka
Pallas's Leaf Warbler	Karafuto-mushikui
Arctic Warbler	Meboso-mushikui
Yellow-browed Warbler	Kimayu-mushikui
Sakhalin Leaf Warbler	Ezo-mushikui
Eastern Crowned Warbler	Sendai-mushikui
Ijima's Leaf Warbler	Iijima-mushikui
Asian Brown Flycatcher	Ko-samebitaki
Dark-sided Flycatcher	Same-bitaki
Grey-streaked Flycatcher	Ezo-bitaki
Yellow-rumped Flycatcher	Mamijiro-kibitaki
Narcissus Flycatcher	Ki-bitaki

English name	Japanese name
Mugimaki Flycatcher	Mugimaki
Taiga Flycatcher	Ojiro-bitaki
Blue-and-white Flycatcher	O-ruri
Japanese Robin	Komadori
Ryukyu Robin	Akahige
Rufous-tailed Robin	Shimagoma
Red-flanked Bluetail	Ruri-bitaki
Bluethroat	Ogawa-komadori
Siberian Blue Robin	Ko-ruri
Siberian Rubythroat	Nogoma
Daurian Redstart	Jo-bitaki
Siberian Stonechat	No-bitaki
Desert Wheatear	Sabaku-hitaki
Japanese Paradise Flycatcher	Sankoucho
Long-tailed Tit	Enaga
Coal Tit	Hi-gara
Willow Tit	Ko-gara
Marsh Tit	Hashibuto-gara
Varied Tit	Yama-gara
Great Tit	Shiju-kara
Chinese Penduline Tit	Tsurisu-gara
Eurasian Nuthatch	Goju-kara
Eurasian Treecreeper	Kibashiri
Japanese White-eye	Mejiro
Bonin Honeyeater	Meguro
Black-naped Oriole	Kourai-uguisu
Tiger Shrike	Chigo-mozu
Bull-headed Shrike	Mozu
Brown Shrike	Aka-mozu
Long-tailed Shrike	Takasago-mozu
Great Grey Shrike	O-mozu
Chinese Grey Shrike	O-karamozu
Black Drongo	Ouchu
Eurasian Jay	Kakesu
Lidth's Jay	Ruri-kakesu
Eurasian Nutcracker	Hoshi-garasu
Common Magpie	Kasasagi
Azure-winged Magpie	Onaga

English name	Japanese name
Common Raven	Watari-garasu
Carrion Crow	Hashiboso-garasu
Large-billed Crow	Hashibuto-garasu
Rook	Miyama-garasu
Daurian Jackdaw	Kokumaru-garasu
Red-billed Starling	Gin-mukudori
Daurian Starling	Shiberia-mukudori
Chestnut-cheeked Starling	Ko-mukudori
White-shouldered Starling	Kara-mukudori
Common Starling	Hoshi-mukudori
White-cheeked Starling	Mukudori
Eurasian Tree Sparrow	Suzume
Russet Sparrow	Nyunai-suzume
Brambling	Atori
Asian Rosy Finch	Hagi-mashiko
Pine Grosbeak	Ginzan-mashiko
Pallas's Rosefinch	O-mashiko
Common Rosefinch	Aka-mashiko
Arctic Redpoll	Ko-beni-hiwa
Common Redpoll	Beni-hiwa
Eurasian Siskin	Ma-hiwa
Oriental Greenfinch	Kawara-hiwa
Common Crossbill	Isuka
Two-barred Crossbill	Naki-isuka
Japanese Grosbeak	Ikaru
Chinese Grosbeak	Ko-ikaru
Eurasian Bullfinch	Uso
Hawfinch	Shime
Long-tailed Rosefinch	Beni-mashiko
Yellowhammer	Ki-aoji
Pine Bunting	Shiraga-hojiro
Meadow Bunting	Hojiro
Japanese Reed Bunting	Ko-jurin
Tristram's Bunting	Shirohara-hojiro
Chestnut-eared Bunting	Hoaka

English name	Japanese name
Little Bunting	Ko-hoaka
Yellow-browed Bunting	Kimayu-hojiro
Rustic Bunting	Kashiradaka
Yellow-throated Bunting	Miyama-hojiro
Yellow-breasted Bunting	Shima-aoji
Chestnut Bunting	Shima-nojiko
Japanese Yellow Bunting	Nojiko
Black-faced Bunting	Aoji
Grey Bunting	Kuroji
Pallas's Reed Bunting	Shiberia-jurin
Reed Bunting	O-jurin
Lapland Bunting	Tsumenaga-hojiro
Snow Bunting	Yuki-hojiro

Tadao Shimba

Black-faced Spoonbill. December, Okinawa-Hontou

INDEX